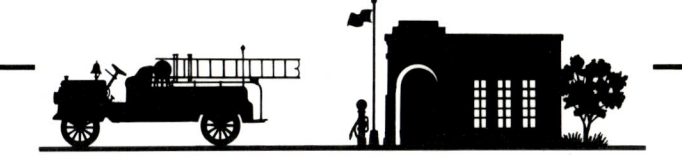

Motorized Fire Apparatus of the West 1900-1960

Wayne Sorensen and Donald F. Wood

**Motorized Fire Apparatus of the West 1900-1960
by Wayne Sorensen and Donald F. Wood**

Copyright © 1991 by Transportation Trails

All Rights Reserved
No part of this book may be reproduced in any manner whatsoever without written permission from the publisher, except in the case of brief quotations embodied in reviews and articles.

For information write to:

**Transportation Trails
9698 West Judson Road
Polo, Illinois 61064
Phone: (815) 946-2341**

First Printing: 1991
Manufactured in the United States of America

Library of Congress Cataloging in Publication Data

Sorensen, Wayne, 1918-
  Motorized Fire Apparatus of the West, 1900-1960 / Wayne Sorensen and Donald F. Wood.
      p.   cm.
  Includes bibliographical references.
  Includes index.
  ISBN 0-933449-11-9
  1. Fire extinction — West (U.S.) — History — 20th century. 2. Fire fighting equipment industry — United States — History — 20th century.
  I. Wood, Donald F., 1935-   II. Title.
  TH9360.S67   1990
  628.9'25'0978 — dc20   90-11010
                                                    CIP

Publisher's Credits

computerized typesetting/layout: Gloria Bellows, National Bus Trader, Inc., Polo, Illinois.
cover art: Terry Newman, The Art Factory, Ltd., Elm Grove, Wisconsin.
printing/binding: Rochelle Printing Co., Rochelle, Illinois.

Table of Contents

	Dedication	4
	Introduction	5
	Acknowledgements	7
1	**Fire Apparatus**	9
	An Introduction to the Development and Use of Fire Apparatus in the West.	
2	**The "Western" Apparatus Builders**	21
	A Review of Individual Fire Apparatus Manufacturers West of the Mississippi River.	
3	**The "Eastern" or "National" Manufacturers**	91
	Prominent Apparatus Manufacturers East of the Mississippi River which Sold Products in the West.	
4	**More Apparatus on Commercial Chassis**	161
	A Pictorial Survey of Commercial Chassis Built by Unknown Manufacturers.	
	Bibliography	227
	Index	229

Dedication

To the memory of
Erik Wayne Sorensen
1953-1979

About The Authors

Wayne Sorensen is a Professor Emeritus at San Jose State University, San Jose, California. He is an historian of old fire apparatus and has photographed fire apparatus for over 50 years. He is a member of the Fire Associates of Santa Clara County (California), the Society for the Preservation and Appreciation of Antique Motor Apparatus in America, the Sacramento Fire Buff Club and the American Truck Historical Society.

Donald F. Wood is a Professor of Transportation at San Francisco State University, San Francisco, California. He is the co-author of several widely-used college-level transportation and logistics texts, and writes about old trucks as a hobby. He belongs to the American Truck Historical Society and to the Society of Automotive Historians.

The co-authors are very interested in fire apparatus manufactured in the West, especially that built in small shops which were not mentioned in this book. They are also attempting to assemble photographs and other materials concerning apparatus used to fight forest and brush fires for use in a book similar to this.

If you have information concerning these rigs and their manufacture, or if there are errors in the text, or if you have any other information you wish to call to the co-authors' attention, please write to: Professor Donald F. Wood, School of Business, San Francisco State University, 1600 Holloway Avenue, San Francisco, California 94132.

Introduction

This book deals with fire apparatus of the West. While many fine books on makes of fire apparatus and fire departments are appearing, the co-authors felt that they would approach the topic from a different perspective. We tried to provide a more balanced look at all the fire apparatus in use, rather than the most elaborate rigs. Also, we looked at the many distinguished fire apparatus manufacturers and outfitters who were — and are — located in the West.

We also recognize that fire fighting is a science, and we have tried to trace the development of apparatus from the start of this century.

Wayne Sorensen
San Jose State University

Donald F. Wood
San Francisco State University

This is a light wagon mounted on a 1952 GMC chassis for use by the San Francisco Fire Department. The lighting equipment had been originally carried on a 1928 Kleiber chassis. CHET BORN, SAN FRANCISCO FIRE DEPARTMENT.

Above: **Dart trucks were built in Waterloo, Iowa until 1925, when the firm moved to Kansas City, Missouri, where it exists today. Now it's part of Paccar Corporation, building large, offroad rigs for use in mining. This picture shows a Dart combination chemical and hose car, from about the time of World War I.** SMITHSONIAN INSTITUTION.

Below: **Shown is a pair of Webbs used by Anaheim, California. On the left is a Webb 900-gpm rotary pumper, powered by a six-cylinder 90-hp engine and with a guaranteed speed of 60 mph. At the right is a Webb combination chemical and hose car with a four-cylinder engine and guaranteed to go at least 50 mph. It carried a 40-gallon chemical tank and 100 feet of 2½-inch hose.** ROBERT SAMS.

6 • *Motorized Fire Apparatus of the West*

Acknowledgements

A number of individuals were of assistance to us. We'd like to thank them and hold them blameless for any errors the book may still contain.

We thank: Dick Adelman; Bob Allen; Phillip S. Baumgarten; Roland Boulet; Chris Cavette, editor of the California chapter newsletter for the Society for the Preservation and Appreciation of Antique Motor Fire Apparatus in America; Clancy Crum; Dean Arthur Cunningham of the School of Business at San Francisco State University; L.N. "Pop" Curtis; Paul Darrell; Ed Gardiner; John G. Graham; Phil Green; Richard A. James; Dan Martin; Dale Magee; Richard S. Nelson of San Francisco State University; Walt Pittman; Chuck Rhoads; Robert Sams; Richard Schneider; Ed Sibert; Charlotte Sorensen; Chester Bailey of Van Pelt; John C. Watson of Kenworth; Bill West; and Leonard Williams.

A Fabco mobile fire extinguisher used by an industrial fire department. FABCO.

COMBINATION CHEMICAL WAGONS
Equipped Complete on Any Chassis

You name the chassis and we fit it up complete with the desired, single tanks, double tanks, hand extinguishers, axes, ladders, bells, landing nets, lanterns, etc., etc.

We are the leading builders of chemical tanks for fire apparatus. Note the illustrations of some of our designs.

We are always pleased to furnish estimates for Tanks or other equipment desired. Write for our Catalogue on Tanks.

Childs Truck (Buick Chassis) in Service at New Milford, Conn.

"Utica" Type "E" 40-Gallon Copper Tank with large opening in the end.

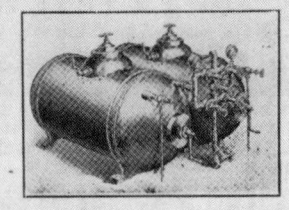

Two "Utica" Type "E-1" Modified Holloway 35-Gallon Copper Tanks, same as Type "E" except it does not have the large cap at the end.

Two "Utica" Type "HH" Holloway 60-Gallon Tanks, complete with By-Pass, 1-inch piping.

"Utica" Type "H-2" Holloway Type Copper Tank. This tank has two 30-Gallon compartments connected with by-pass valves and piping so that either compartment can be operated independently.

Two "Utica" Holloway Type "H" 40-Gallon Tanks, complete with 2½-inch water hose connection and necessary By-Pass valves ¾-inch piping.

Childs Chemical (Ford Chassis).

Windsor, Ont., June 12, 1916.
O. J. Childs Co., Utica, N. Y.

Gentlemen: I am pleased to inform you that we have recommended your type H "Utica" Holloway tanks for our new motor apparatus. We have three pieces of your equipment in service in our Department at the present time, and find them entirely satisfactory both in regards to service and mechanical construction. I take great pleasure in recommending your chemical apparatus where the requirements call for heavy fire service. I am, very truly yours,

(Signed) C. J. DeFIELDS,
Chief of Fire Dept.

"Utica" Type "H" Holloway 35-Gallon Copper Tank.

Childs Truck (Selden Chassis) in Service at Brockport, N. Y.

O. J. CHILDS CO. - Utica, N. Y., U. S. A.
Manufacturers Chemical Fire Apparatus.

1

Fire Apparatus

This book has a regional orientation. Today, when one thinks about fire apparatus, it is hard to think of regional differences. Years ago this was not so. There were many hundreds of truck manufacturers (most serving little more than a local or regional market). The same held for body builders, many were one-time wagon or carriage builders, or even blacksmith shops.

Trucks built in the West were different. The terrain in which they were used was more rugged than in other parts of the country. Distances between major cities were greater, and double bottom truck/trailer combinations were allowed to flourish. Hence, Western trucks had to be more powerful. From a nationwide standpoint, we tend to think of the more powerful trucks being developed in the West and gradually spreading to the Midwest and South and East as interstate trucking began developing to its full potential.

It's harder to think that fire apparatus used in the West was much different than that used elsewhere. Urban and suburban developments in the West tended to be more spread out; a local fire chief might have to locate his equipment over a wider geographic area than would his counterpart in the East. A lack of urban concentration might also mean fewer ladder trucks. Streets and highways in the West were wider, traffic congestion and poor maneuverability were probably less of a concern. Much of the development in the West is newer than that elsewhere in the nation. Decisions concerning apparatus and station siting were more likely to be made in the motorized era.

Water supplies in the West always have been more of a problem than elsewhere, although this cannot be translated directly into fire apparatus needs since often the problem was addressed first by those responsible for providing municipal water supplies. One of the lessons from the 1906 and 1989 San Francisco earthquakes was the vulnerability of a city's water supply system. Related to this, we do believe that the use of fire trucks with large (for their time) water tanks developed in the West. (As, on the other side of the same coin, did the John Bean high-pressure pump, which made very good use of a limited water supply.)

One type of fire, unique to the West, is the large-scale forest or brush fire. Every summer, one reads of forest fires in remote areas which are difficult to reach and to contain. Often, these fires sweep close to developed areas and destroy homes.

Several West Coast manufacturers also developed major apparatus components which were, in fact, used nationwide. They will be discussed in the following chapter. The two outstanding examples are the Hall-Scott engine and the Gorham pump.

This book gives more emphasis to chassis and fire apparatus builders who were — or are — located in the West. ("West" is interpreted loosely, and in two ways. In Chapter Two, when discussing the manufacturers and assemblers of apparatus, we deal with the area west of the Mississippi River, although it is evident that our knowledge of manufacturers in the "far" West is more complete. In Chapters Three and Four, which deal more with users of apparatus, we — for the most part — restrict ourselves to the area west of the Rockies.)

Luckily for the truck or fire apparatus historian, many old rigs exist. This is because the trucks were kept and maintained inside buildings for the first 30 to 40 years of their life. Many fire departments — or groups interested in the preservation of old apparatus — keep one or more rigs for use in parades or at "musters," which are fun activities sponsored by fire fighting buffs.

Fire apparatus itself is much older than the motor truck. Pumps, chemical and hose carts and even aerial ladder trucks were in existence before the turn of this

---

PRECEDING PAGE

This ad, from about 1917, is for chemical tanks manufactured by the O.J. Childs Company of Utica, New York. Note that they could be installed as single units, although pairs were more common because one could be recharged while the other was "working." Note also that they could be fitted into conventional truck beds. In this ad, the three truck chassis are, from top, Buick, Ford and Selden.

The earliest self-propelled fire apparatus depended upon steam. This rig was built just after the turn of this century by the firm which was to become American-LaFrance. It was powered by two two-cylinder, four-cycle Mason motors. This one went to New London, Connecticut. Only a few were built. GUS A. JOHNSON.

century and were pulled by horses. Indeed, the first adaptation of motor trucks to fire fighting was to use the motor truck as a tractor, taking the place of the horses. Only later was the truck's power plant utilized to pump water, lift ladders or generate electric power for floodlights.

The first motorized vehicles used by fire departments were chiefs' cars, still sometimes referred to as "buggies." The auto's speed was superior to that of the horse. Some chief's cars were outfitted with light fire fighting and first aid equipment.

For readers unfamiliar with fire apparatus types, we list trucks by their various functions. Most trucks are outfitted to perform more than one function.

Pumpers are the most common, and are the only apparatus correctly referred to as fire engines. The use of the gasoline engine to pump water was probably as significant as the use of the gasoline engine to pro-

Here is another early self-propelled fire engine powered by steam. This 1908 Amoskeag was purchased by Vancouver, British Columbia. It weighed eight tons and was rated as being able to pump 1,200 Imperial gallons a minute. Its top speed was 12 mph. GUS A. JOHNSON.

While not a West Coast product, this picture of a 1909 Luitweiler is included because it shows the early use of a single engine to drive both the pump and truck. Note the large flywheel in center, below the hose basket. This machine was produced by the Luitweiler Pump and Engineering Company of Rochester, New York. WESTERN RESERVE HISTORICAL SOCIETY, CLEVELAND, OHIO.

pel the fire apparatus. For a short time, some gasoline-powered pumpers were pulled by horses. (Pumpers are positioned close to hydrants or other water sources and pump water toward the direction of the fire.)

Here, from a 1926 book by Robert McNeish, entitled *The Automobile Fire Apparatus Operator*, is the list of tasks to be performed by the operator of a Seagrave pumper:

"Spot engine at hydrant with engine running.
"Set emergency brake.
"Connect suction, noticing if wire strainer is in suction.
"Open hydrant.
"Connect lines of hose to pump discharge.
"Open discharge gates.
"Throw out clutch with hand lever.
"Mesh pump gears.
"Let in clutch.
"Open throttle until pump discharge gauge shows 10 pounds above desired pressure.
"Open valve in water pressure regulator line and adjust water pressure regulator to desired pressure.
"Examine valve on auxiliary cooler; open if it is closed.
"Turn down grease cup on rear pump bearing two turns every hour.
"Oil valve stems; repeat every hour.
"Notice oil level in crankcase; put in more oil when needed.
"Notice pressure gauges; if hands jump, partly close cocks next to gauges.
"Examine radiator; fill up if necessary.

"Inspect all piping, hose connections, etc., for leaks; tighten connections, if necessary, or make a note of them so that faults can be corrected at first opportunity.
"There is nothing more to be done except to watch operation of apparatus and make any adjustments that are required.
"Replenish supply of gasoline, oil and water whenever needed."

In cities, pumpers were accompanied by hose wagons. The hose wagon might be equipped with a deck or deluge gun, through which water from several hose lines would be played onto the fire.

Another form of apparatus was the chemical car. Chemical cars carried two or three large soda/acid fire extinguishers, which would be brought into play, one at a time, and would feed through a single hose onto the fire. Chemical cars and hose wagons might be combined, and these rigs were known as combination hose and chemical wagons. Eventually small "booster" water tanks and pumps replaced the chemical extinguishers.

Here, from a 1921 book, Victor Page's *The Modern Motor Truck*, is the description of the equipment carried on a combination double-chemical tank/hose body, which could carry 1,000 feet of 2½-inch hose:

"Electric starter; two electric head, two dash, one swivel search and one tail light; two 35-gallon nickel-plated chemical tanks with 200 feet of three-quarter-inch chemical hose on reel; one 20-foot extension ladder; one roof ladder; two fire lanterns; one pick back-fire axe; one crowbar; two tool boxes;

Engine Company 14 in Oakland, California, used this horse-drawn, gasoline-powered Waterous pump. In the rear is a hose cart, pulled by a single horse. Photo is from 1910. PAUL DARRELL.

In 1904, Turlock, California, purchased this Howe horse-drawn gasoline pumping engine. In 1923, they sold the rig to neighboring Atwater. The rig has a two-cylinder pancake engine and still performs at musters. WAYNE SORENSEN.

The Waterous Engine Works of St. Paul, Minnesota, built this horse-drawn 350-gpm gasoline-powered pumper for Fairfield, California, in 1909. It remained in service until 1931 and still can be seen at musters. WAYNE SORENSEN.

12 • *Motorized Fire Apparatus of the West*

The use of water tanks and small pumps was popular in the West. Here's an early rig, with a rectangular tank on a flatbed truck, fighting a brush fire. CALIFORNIA STATE ARCHIVES.

A piston pump mounted at the front of an Ahrens-Fox chassis. RICHARD HENRICH COLLECTION.

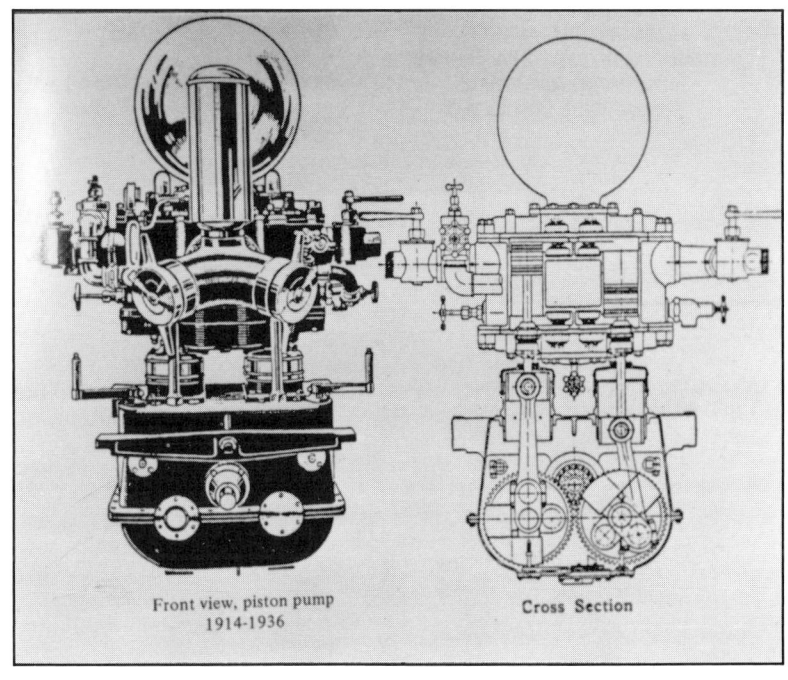

These two cutaway drawings are an Ahrens-Fox piston pump. These pumps originated on steam apparatus where they could be directly integrated to the steam engine's back and forth movement, i.e. through direct connection the stroke of the steam engine's piston was the same as that of the pump. The pump pistons are below the globe and — in being driven upward — create pressure. Piston fire pumps had two disadvantages: They were more difficult to link with an internal combustion engine's rotary motion, and they had more parts than other types of pumps.

Front view, piston pump 1914-1936   Cross Section

Motorized Fire Apparatus of the West • 13

Here is a rotary gear pump mounted on a truck chassis. Rotary pumps were very popular during the World War I era. One source indicated that of the fire apparatus pumps built in the 1910-21 period, 2,944 were rotary gear pumps, 598 were centrifugal pumps and 395 were piston pumps.

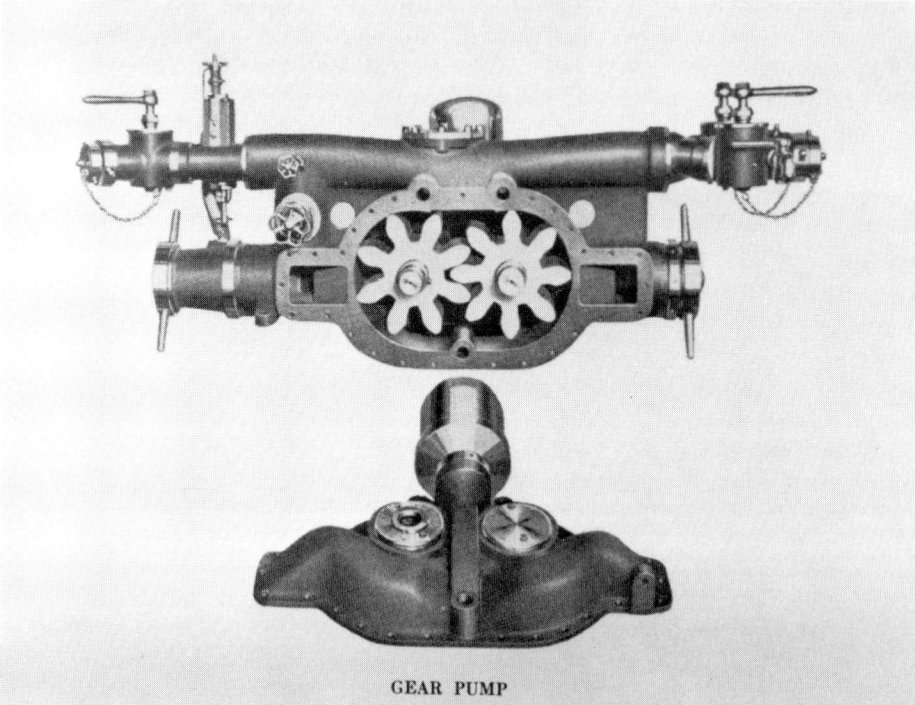

An American-LaFrance rotary gear pump.

GEAR PUMP

This shows in more detail the flow of water through the rotary gear pump. The left gear moves clockwise, the right gear counterclockwise. The water is discharged from outlets on the top. All gears are finely fitted and, as they rotate, they create a vacuum. The gears and case are of cast bronze. The pump has very few parts. Its main limitation is that when drafting from ponds or streams, gravel or sand passing through the pump can damage pump vanes.

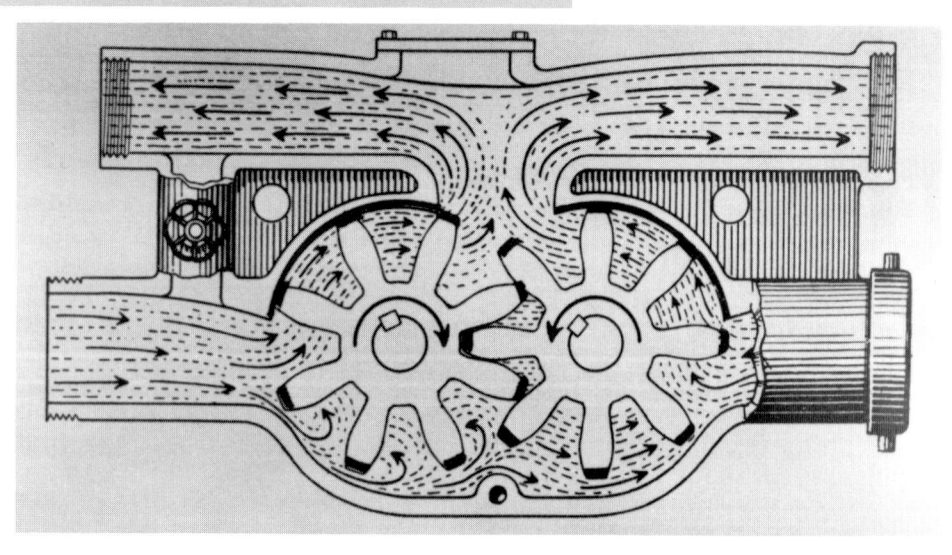

14 • *Motorized Fire Apparatus of the West*

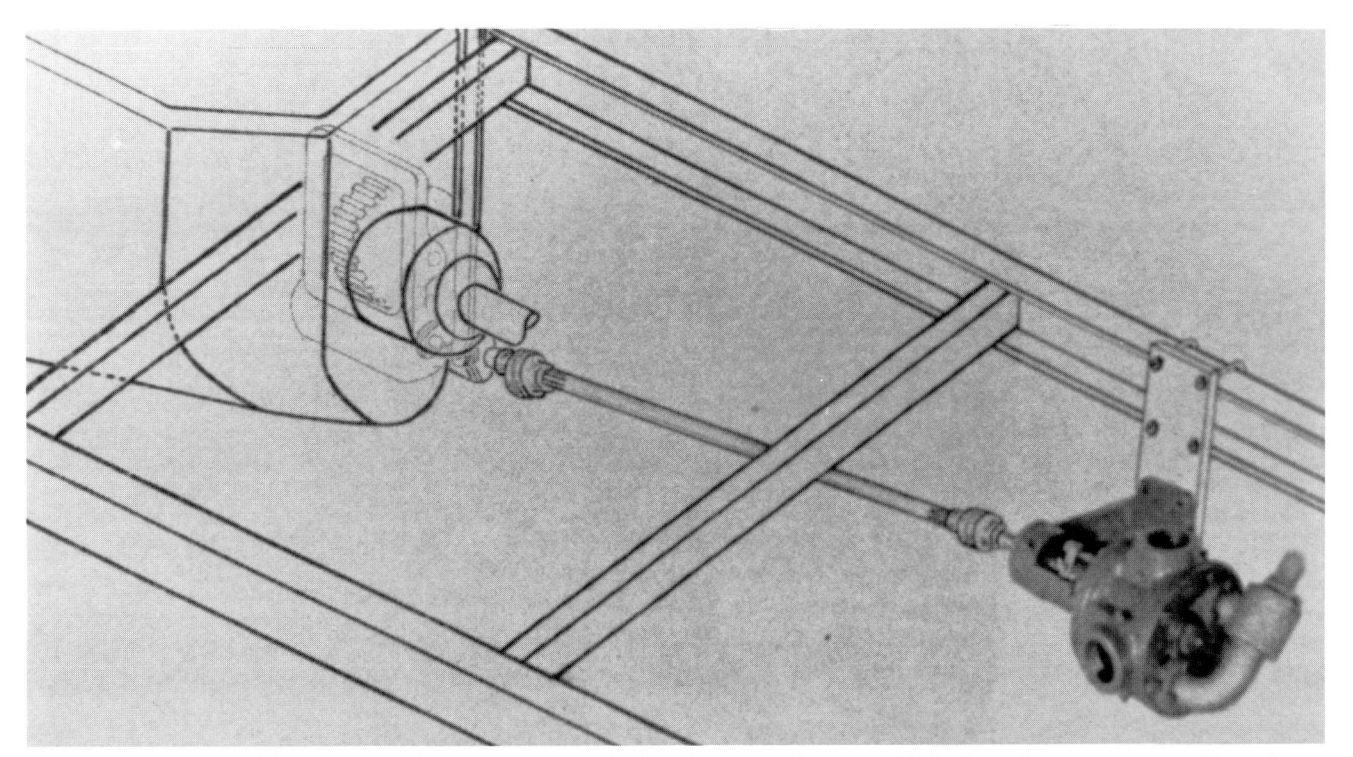

This picture appeared in a 1938 catalog that contained advertisements from outside body and accessory suppliers, showing equipment for Chevrolet trucks. The ad for "Viking Fire Truck Pump Units," manufactured by the Viking Pump Company of Cedar Falls, Iowa, indicated three capacities: 35, 50 and 90 gpm, which meant they were intended for booster pumps or rural fire fighting. This picture, one of several, shows how the pump can be connected to the power take-off and mounted on the truck's frame. An individual contemplating outfitting fire apparatus would refer to ads such as this as he planned the layout of various features. HARRAH'S AUTOMOBILE COLLECTION, RENO, NEVADA.

two play pipe holders; one locomotive bell, whistle on exhaust; one shut-off nozzle; two nozzle tips; two hose spanners; one soda bag; two acid receptacles; and receptacle holder/filling cap wrench."

Ladder trucks carried ladders of varying lengths. Aerial ladder trucks were trucks equipped with an aerial ladder, powered by either mechanical or hydraulic means.

City service trucks carried ladders and a variety of other rescue equipment. Victor Page's 1921 book gives this list of equipment for a city service truck:

"One 50-foot rapid hoist rope and pulley extension (ladder) to extend 50 feet with supporting poles; one 35-foot rapid hoist rope and extension (ladder) to extend 35 feet; one 28-foot single ladder; one 24-foot single ladder; one 20-foot single ladder; one 16-foot roof ladder; one 12-foot roof ladder; two 25-foot single side ladders; . . . two Crotch poles; four fire department axes; six pike poles—assorted lengths; two wall picks; two crowbars; two shovels; one wire cutter; one door opener; one tin roof cutter; two pitchforks; two No. 2 Babcock fire extinguishers; four fire department lanterns; one wire basket 14" x 30" x 6' in rear of driver's seat for tools, etc.; one tool box."

Nearly all apparatus are combinations or variants of the types just listed. A few other, very specialized rigs would be floodlight trucks, water towers, tankers, brush rigs, etc. Some of these are illustrated in this book.

After discussing fire apparatus body types, we also examine the chassis suppliers and the apparatus body builders. Chassis suppliers built or assembled the truck's chassis. This would usually consist of the frame, axles, power plant and transmission. Often the front metalwork as far back as the cowl, windshield or cab would be included. In this state, the chassis would be driven (or shipped by rail) to a body builder's shop where the remainder of the truck would be added. For fire engines, the first step was to install the water pump, either midship or sometimes in the front, forward of the radiator. A booster tank and booster hose reel would then be added. Finally, sides would be added which would enclose the hosebed. Racks for suction hose and ladders, hooks, equipment cabinets, lights, etc., would all be added. This was the way most apparatus was — and is — built, directly on the easily recognizable makes of "commercial" chassis: Fords, Chevrolets, Dodges, Reos, Diamond-Ts, etc. This book devotes considerable space to apparatus built on commercial chassis, and this is for several reasons. In terms of numbers, they were — and are — the most widely used. They are the backbone of many smaller and medium-sized departments. Also, other books on fire apparatus tend to neglect them.

Custom apparatus builders built the entire truck. These are the well-known names: Ahrens-Fox, American-LaFrance, Seagrave, etc. (Although these custom builders would also outfit commercial chassis as outlined in the previous paragraph.) Custom apparatus was more expensive, and usually associated with big-city, full-time fire departments.

Actually, the term custom is somewhat of a misnomer since almost no apparatus was built "on spec," i.e. in advance of a specific order. That is because each fire chief would want to incorporate a

Motorized Fire Apparatus of the West • 15

A view of an enclosed centrifugal pump.

CENTRIFUGAL PUMP—REAR VIEW

This is a cutaway drawing of a centrifugal pump. Rapid rotation forces the water outward to a discharge pipe. Sand and gravel cause little damage, and the pump can tolerate sudden changes in engine speed or water flow.

A midship mounting of a centrifugal pump, but they also can be mounted in front of the truck's engine. Centrifugal pumps are the most commonly used fire pumps today.
AMERICAN-LaFRANCE.

16 • *Motorized Fire Apparatus of the West*

Three 1956 Fords outfitted by The Boardman Company with 750-gpm pumps. THE BOARDMAN COMPANY.

Alameda, California, was one of the first West Coast fire departments to motorize. Shown are the chief's car, a 1912 Paige; a 1916 Packard hose car; a 1907 Waterous 600-gpm pumper; and a 1914 Knox-Martin tractor pulling a city service truck. PAUL DARRELL.

Here is the early motorized apparatus in the Berkeley (California) Fire Department's Fire Station No. 2. On the left is a 1915 Boyd city service truck; at the right a 1915 Locomobile used as the chief's car. Behind it is a 1915 Seagrave-Gorham pumping engine and behind that a 1912 Webb hose wagon. BERKELEY FIRE DEPARTMENT.

This photo, from the mid-1920s, shows Logan, Utah, motorized apparatus. From the left are a 1923 American-LaFrance 600-gpm pumper, a 1913 American-LaFrance 600-gpm pumper, a 1925 Dodge chemical car, and a 1924 Type 36 LaFrance-Brockway Torpedo chemical and hose car. LOGAN FIRE DEPARTMENT.

few features which he thought would be of particular value to his operation.

One other source of apparatus bodies was shops. Large fire departments had their own shops where they would modify and repair equipment, and sometimes build bodies for special fire fighting needs. In some small communities, apparatus was actually outfitted in local repair shops, where a pump (powered by the truck's power take-off), a water tank, hose reels and racks, searchlights, etc., would all be installed.

In addition to apparatus builders, there were also well-known manufacturers of specific components, the best example being pumps. Other examples were manufacturers of sirens and warning lights and aerial ladders.

The order of presentation for this book is first by apparatus builder. For those that made custom equipment, the custom equipment is shown first, then the apparatus mounted on commercial chassis. Chapter Two deals with apparatus builders located in the West. Chapter Three covers builders located in the East who sold in nationwide markets. The last chapter of the book is devoted to apparatus on commercial chassis only. In that part of the book, relatively little is known about the apparatus outfitter. Within these breakdowns, apparatus is pictured in approximate alphabetical and chronological order. □

Apparatus lined up in front of Pocatello (Idaho) Fire Department's Fire Station No. 1 just before World War II. In the distance a 1937 Seagrave 65-foot aerial ladder truck, a 1915 American-LaFrance Type 12 pump and hose car, a 1923 American-LaFrance combination chemical and hose car and a 1933 Graham-Paige auto equipped with a booster tank and used as a chief's car. POCATELLO FIRE DEPARTMENT.

Motorized Fire Apparatus of the West • 19

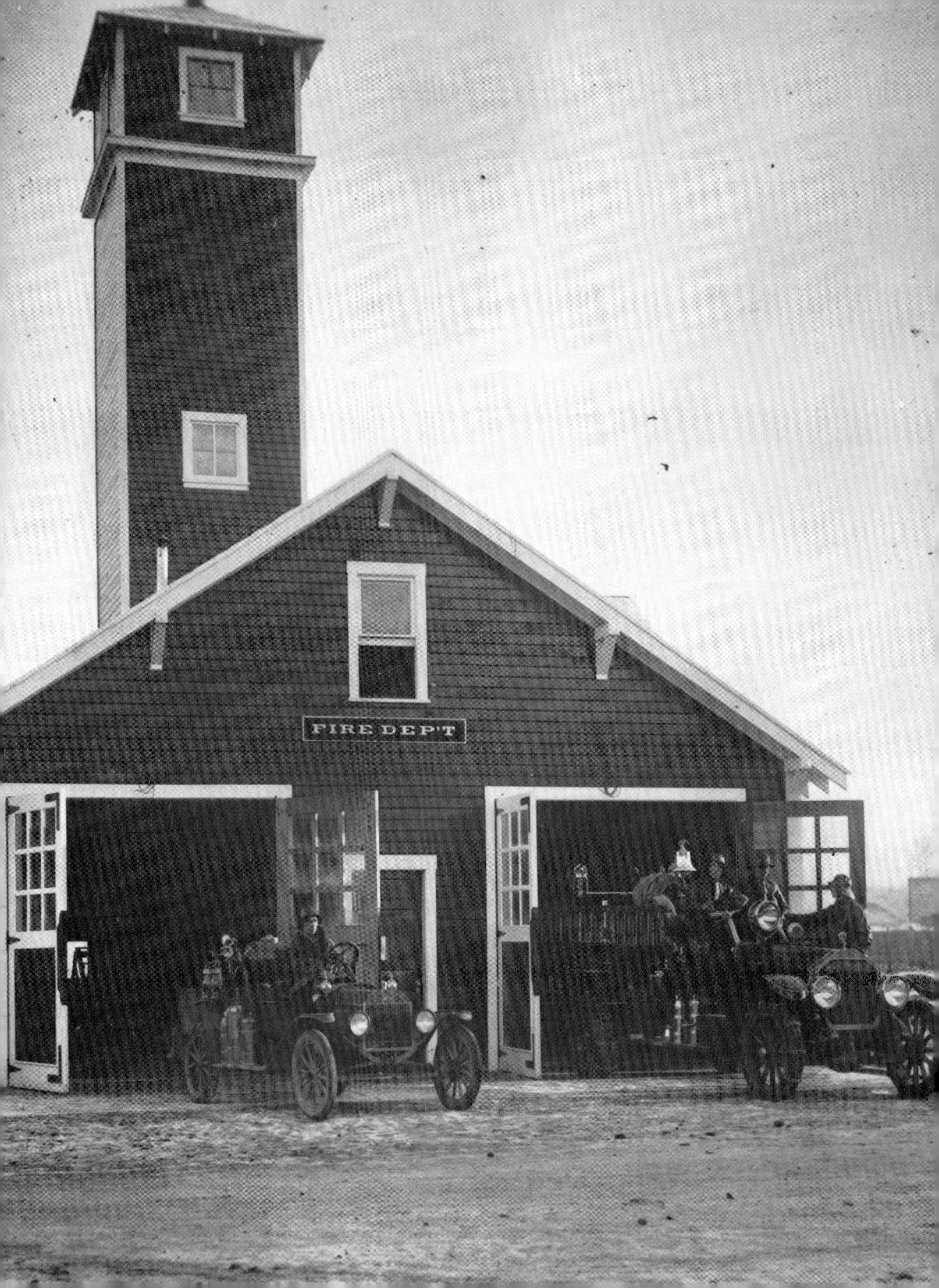
FIRE DEP'T

2

The "Western" Apparatus Builders

This chapter deals with apparatus manufactured in the West, with the area being defined approximately as that west of the Mississippi River. We have included manufacturers whose operations may have moved in and out of the area during the course of their existence. Note also that over time, firm names may change.

It is difficult to develop an "airtight" classification system into which all apparatus will fit. Often, several well-known names are associated with a single rig: the chassis manufacturer, the apparatus body manufacturer and the pump or aerial ladder manufacturers. Sometime, during the course of the vehicle's life, it will be rebuilt, repowered and re-equipped with components from even different manufacturers than participated in the earlier assembly. Apparatus also has a very long life, and the same rig's appearance often changes. A common change which took place in the 1920s was the substitution of pneumatic for solid tires; in the 1940s, sealed-beam headlights were substituted for the much more attractive — but less effective — original drum-type units.

American Car & Foundry Company
Berkeley, California

In 1925, the American Car & Foundry Company (whose product was called the "ACF") of Detroit, purchased the Fageol Truck & Coach Company of Oakland, California, which had been operated by the Fageol brothers, two well-known West Coast truck and bus builders. The Fageol brothers became vice presidents of ACF, but the small Oakland plant remained relatively independent. (The Fageol's truck

Above: This is a 1915 Boyd city service truck originally built for the Berkeley (California) Fire Department by James Boyd & Brothers, Inc. of Philadelphia, a firm founded in 1819. The vehicle was rebuilt by the American Car & Foundry's Hall-Scott plant and repowered with a four-cylinder Hall-Scott engine. PAUL DARRELL.

Preceding Page: This picture, taken about the time of World War I, illustrates some of the differences between "custom" apparatus and apparatus mounted on commercial chassis. At the right is a custom American-LaFrance pumper, while at left is a smaller Ford Model T chemical car. The two rigs belonged to the Alaskan Railroad Fire Department and were stationed on Ship Creek in Anchorage. ANCHORAGE HISTORICAL AND FINE ARTS MUSEUM.

Motorized Fire Apparatus of the West • 21

The Berkeley (California) Fire Department had this 1914 Knox chemical and hose wagon rebuilt by American Car & Foundry into a high-pressure hose car. A new electrical system, hood, radiator and Hall-Scott motor were installed. PAUL DARRELL.

This combination chemical and fire hose car was originally built in the shops of the Berkeley (California) Fire Department, and later rebuilt by American Car & Foundry. PAUL DARRELL.

This 1914 Knox chemical and hose car was used by the Berkeley (California) Fire Department. In the late 1930s American Car & Foundry rebuilt it. PAUL DARRELL.

22 • Motorized Fire Apparatus of the West

This American-LaFrance Type 10 chassis (serial number 552) was owned by the Berkeley (California) Fire Department, whose shops installed the hose body and chemical tanks. In the 1930s, the rig was rebuilt by American Car & Foundry and received a new Hall-Scott motor. It remained in service until the 1950s. JOHN G. GRAHAM.

operation encountered financial difficulties in both the late 1920s and 1930s, and it went out of business just before World War II; the plant and many materials were sold to T.A. Peterman who began building his Peterbilt trucks there.)

American Car & Foundry had a second plant in Berkeley as well, located at 2850 Seventh Street. Here the Hall-Scott Motor Car Company was located. The Hall-Scott firm had been founded by Elbert Hall and Bert Scott in about 1908 to build racing cars. They also built a gasoline-powered railroad engine for moving light railroad cars. Soon, they began building aircraft engines and Hall is credited as being one of the designers of the famous World War I Liberty aircraft engines. Soon, Hall-Scott supplied engines for Fageol buses and trucks and for other motor vehicles (this was still during the time when there were numerous auto and truck manufacturers who would "assemble" their vehicles from components purchased completely from outside suppliers).

American Car & Foundry (in conjunction with J.G. Brill Company of Philadelphia) purchased Hall-Scott in 1925, and a corner of the Hall-Scott plant was reserved for the rebuilding and repowering of fire apparatus. Repowering was especially necessary for pumpers, because the engine often would receive much more wear than the chassis.

The ACF name disappeared in the early 1950s. The Hall-Scott firm survived as a subsidiary of other firms until about 1960. It was handicapped because it had not developed a diesel engine, which became increasingly popular for heavy trucks and buses after World War II.

Re-powered With A Hall-Scott Model 400

This is part of a page from an eight-page Hall-Scott brochure from the late 1930s, shows a 1930 Mack repowered with a Hall-Scott engine.

Motorized Fire Apparatus of the West • 23

Famous BARTON-AMERICAN PUMPS

Your Choice of Front-Mounted or Midship Types

...manufactured by our affiliate, American Fire Pump Company

The 500 G.P.M. and 750 G.P.M. Front-mounted Pumps combine finest quality materials and workmanship with simplicity that reduces construction and maintenance costs. All controls are located right at the pump.

Effective **frost-proof design** gives you positive operation in coldest weather and in summer keeps engine **cooler** — a year 'round advantage!

Exclusive positive-action pump clutch has interlocking dogs that **can't** slip, can't fail at a fire. Universal joint shaft drive from crankshaft. Gear ratios vary to properly suit your engine. Capacities to and including 750 G.P.M., pressures to 400 P.S.I.

Only American offers you the assurance of dependability resulting from over 35 years' successful experience in front pump installation in all makes of trucks.

The Duplex Multi-stage Midship Pump offers extra flexibility and highest performance as a result of its excellent engineering.

In **one** unit you get **two** impellers — **independently shafted and geared.** They run separately, if desired, or together in series. A single lever on the selective gear transmission gives a choice of high volume at normal pressure — normal high pressures — and extremely high pressures for fog guns, **all** at moderate engine speeds. High volume pump handles dirty water without damage to high pressure pump.

One point lubrication — all bearings run in oil inside **one** automotive-type housing. All valves operate **automatically** by pressure — no separate controls. Capacities to and including 1000 G.P.M., pressures to 400 P.S.I., two-stage — 850 P.S.I. on four-stage.

A Barton-American advertisement from the late 1950s, showing both front-mounted and midship types of pumps. One advantage of the front-mount pump was that it gave more room on the truck to carry water or hose.

A 1922 Ford Model T shown with a Barton pump. Note hand-operated siren. AMERICAN FIRE APPARATUS COMPANY.

American Fire Apparatus Company
Battle Creek, Michigan, and Marshalltown, Iowa

Advertisements for this company indicated that it had operations in both Battle Creek, Michigan, and Marshalltown, Iowa. The American Steam Pump Company of Battle Creek manufactured front-mounted pumps for the Barton Products Company of Jackson, Michigan. In the mid-'20s, there was a large market for front-mounted pumps which could be placed on Ford Model T chassis. A simple, centrifugal design was used, and for the Ford Model T, the usual capacity was 250 gallons per minute. For a time, Barton virtually "cornered" the market for low-cost fire engines.

The firm became known as American-Marsh Pumps, and then the American Fire Apparatus Company, with its best-known product being the front-mounted Barton or Barton-American fire engine pump. Available in capacities up to 600 gpm, these pumps adorned the fronts of many Dodge, Chevrolet, Ford, International and other commercial-chassis fire apparatus used in small towns throughout the United States.

The front-mounted pump had many advantages. It was probably the easiest to install, and saved weight and body space, allowing room for a larger booster

A Barton G-100 pump is mounted on the front of a 1936 Chevrolet. Barton pumps were positive-action with a universal joint connected to a crankshaft. All controls are to the right of the pump. Valves opened automatically by pressure, and there were no separate controls. Front-mount pumps were useful in rural areas where water had to be drawn from ponds or irrigation ditches because it was easier to "spot" the pumper. (Note how the grille has been cut away. This was done to give the truck a better weight balance and to support the pump more firmly in the truck's frame). FOREST SERVICE, U.S.D.A.

AMERICAN FIRE APPARATUS CO.
BATTLE CREEK, MICHIGAN
MARSHALLTOWN, IOWA

The Emblem of Quality

Motorized Fire Apparatus of the West • 25

Gaston, Oregon, used this 1938 Ford pumper that was outfitted by a local machine shop after the chassis had seen more than 100,000 miles of service carrying wood. It had a front-mounted Barton 500-gpm pump. Barton pumps were popular in the West, in part because they gave positive action in the coldest of weather. WAYNE SORENSEN.

This 1942 Dodge was used by the Pleasant Valley (California) Fire Department, and had been outfitted by American Fire Apparatus. Gear on running board is Indian back pack tanks firemen carry to fight brush fires. WAYNE SORENSEN.

This is a drawing of a late 1950s Chevrolet 1½-ton chassis carrying a Central pump.

26 • *Motorized Fire Apparatus of the West*

Locomobile Company, of Bridgeport, Connecticut, built this touring car chassis that Ollie Hirst turned into his first fire engine, which was made for Placerville, California. This was a small tank truck. Note the hose reel in the rear. An Evinrude pump is shown on running board. Hirst is the driver. HIRST FAMILY.

tank or hose compartment. The location in the front helped with weight distribution and made it easy for the driver to align the pump with the hydrant or other water source. The front-mounted pump could operate while the truck was in motion. American offered a choice of front-mounted or midships pumps. On all of its pumps, pressure and engine speed were controlled by a governor, even when discharge nozzles were shut off completely.

Typically, the chassis would go from the chassis manufacturer to the American plant, where the pump would be installed. From that point it would be shipped to either an apparatus builder or to a local truck body-building shop near the location of the purchasing fire department. (American accepted the responsibility for the truck chassis from the time it was received at their plant until it was delivered to the customer.)

It exists today as American Fire Apparatus Co., Div. Collins Industries, Inc., Hutchinson, Kansas.

The Boardman Company
Oklahoma City, Oklahoma

This firm began manufacturing fire apparatus for placement onto commercial chassis in 1950. Ford is one of the makes they often use, although they also utilize Dodge, GMC and International chassis.

The firm operates today.

Central Fire Truck Corporation
Manchester, Missouri

This firm produced apparatus in the post-World War II period, with some of its workers being those which General Fire Truck Corporation (see below) left behind. The firm provided pumpers with a range of capacities mounted on Chevrolet and other chassis. Their motto was: "What's under the paint job is what counts." They apparently produced fire apparatus until the mid-1960s; the firm is no longer in business.

Ollie Hirst used Pierce-Arrow chassis for many of his early rigs. The booster and hose car, built in the 1920s, was used at Lake Tahoe (California). Note the coil of hose near the spare tire. HIRST FAMILY.

Motorized Fire Apparatus of the West • 27

This 1930 Fageol-Challenger was used as a demonstrator. It came with a four-cylinder Hall-Scott engine, Fageol's own transmission and worm-gear axle. This rig was especially good at climbing hills, no small asset in much of California. Note the spare tire and cover. Most fire apparatus did not carry spares. HIRST FAMILY.

Initially built for the East Bay (California) Municipal Utility District to protect the area in the hills above Oakland, this 1934 Fageol-Challenger is popular at California musters. It is now owned by Bill Fageol, grandson of the founder of Fageol Motors. The rig has a 200-gpm Viking pump and a 450-gallon water tank. The vehicle is one of the few Fageols powered by a Waukesha engine. WAYNE SORENSEN.

Challenger outfitted this 1937 Chevrolet for use by the Truckee (California) Fire Department, which still owns it. The pump is rated at 500 gpm. Chevrolet chassis were popular with Challenger buyers. BOB ALLEN.

This 1937 Challenger was used by the Quincy (California) Fire Department. Note the streamlining, enclosed equipment compartment and overhead ladders. HIRST FAMILY.

Challenger Fire Equipment Company
Sacramento, California

Ollie N. Hirst was a San Francisco fireman who was assigned as the chief's driver at the time of the 1906 San Francisco earthquake. Sometime later in his life, he relocated about 100 miles to the east, to a town called Placerville, California, near Sacramento and in the western foothills of the Sierra Nevada Range. After watching Placerville's volunteer firemen struggle as they pulled hose carts up the hills, Hirst decided to build his community a piece of motorized fire apparatus. The first fire truck he built was on a Locomobile chassis, adding an Evinrude pump on the running board, a water tank and some hose.

Soon, he was building apparatus for volunteer fire fighters in other nearby communities. He switched to using Pierce-Arrow chassis for building rigs carrying a booster tank and hose. Business prospered and, in 1932, he moved his shop to Franklin Boulevard in Sacramento and used the name ''Challenger'' for his product. Business continued to grow (note that this was during the Depression) and, in 1936, he moved to an even larger site in North Sacramento. He frequently used Fageol chassis, and, in 1939, used the very first Peterbilt to build a fire engine for Centerville (now Fremont), California.

During World War II, the firm went out of business, and Hirst became the fire chief for the Sacramento Army Depot. He died in 1948.

Hagginwood (now a part of Sacramento), California, used this 1938 Fageol-Challenger. It had a Hall-Scott 177 engine and 500- gpm pump. Ollie Hirst is at the wheel. HIRST FAMILY.

Motorized Fire Apparatus of the West • 29

Above: Frequently appearing in local parades with the banner "First Peterbilt," this 1939 Challenger, which looks as though at least one of its parents was a Diamond T, was built by Ollie Hirst for what is now Fremont, California. The Peterbilt chassis is numbered S-100 and the power plant is a Hall-Scott 147. WAYNE SORENSEN.

Below: North Sacramento, California, bought this 1940 Challenger, mounted on a White chassis. It carried a 400-gallon water tank and a 500-gpm pump. Note the fender skirts. HIRST FAMILY.

30 • *Motorized Fire Apparatus of the West*

Hagginwood, California, used this 1942 Challenger rescue squad truck, mounted on a Diamond T chassis. Note ample equipment compartments, floodlights and overhead ladder rack. HIRST FAMILY.

Del Paso-Robla (California) Fire District (now part of Sacramento) used this 1942 Challenger on a Peterbilt chassis (No. S-105). It had a Hall-Scott 160 motor, a Hale pump and a 400-gallon booster pump. PAUL DARRELL.

Challenger used Peterbilt chassis No. S-106 to build this 600-gpm pumper for the Ashland Fire District (near Oakland, California, now the Eden Consolidated Fire Protection District). Hale made the pump and the power plant was a Hall-Scott 147. Note the shape of the door. BOB ALLEN.

Motorized Fire Apparatus of the West • 31

Auburn, California, ran this 1934 Ford Challenger. It had a 105-gallon water tank and two Waterous pumps, one rated at 200-gpm, the other at 400-gpm. This vehicle is now owned by the Citrus Heights (California) Fire Department and has been restored. WAYNE SORENSEN.

This 1940 Autocar-Challenger tank wagon, built for Crockett, California, has a 200-gpm pump, a 500-gallon tank and a 600-gpm Morse turret. Note the equipment compartments and fender skirts. BOB ALLEN.

Two Federal chassis were purchased by the Oakland (California) Fire Department in 1944, and were converted into fire fighting apparatus by Coast after World War II ended. These units had booster tanks and 500-gpm Waterous pumps. WAYNE SORENSEN.

32 • *Motorized Fire Apparatus of the West*

Coast outfitted this 1948 International KB for Tremonton, Utah. It had a 750-gpm Waterous pump and a 450-gallon booster tank. WAYNE SORENSEN.

Coast Apparatus, Inc.
Martinez, California

This firm was started in 1946 by David Barklow and D.B. Bardell when they purchased the fire engine portions of FABCO's business. (FABCO is discussed later.) Both men were former employees of FABCO.

Coast built "custom" apparatus which were used throughout the West. A few were even exported to Arabia. In addition to custom work, Coast outfitted apparatus on chassis provided by Peterbilt, Ford, Chevrolet, Sterling and White. Los Angeles County was a large customer for their rigs built on International chassis.

The Oakland (California) Fire Department used this 1949 Peterbilt-Coast as Engine 18. It was powered by a Hall-Scott 480 engine and had a 1,250-gpm Waterous pump. Also it had a 250-gallon water tank. WAYNE SORENSEN.

Motorized Fire Apparatus of the West • 33

Sterling Motor Truck Company, of Milwaukee, Wisconsin, built a heavy-duty truck that was surprisingly popular on the West Coast. Coast Apparatus outfitted this rig for use at Napa, California. It was powered by a Waukesha motor and had a 1,000-gpm Waterous pump. WAYNE SORENSEN.

This 1949 White- "Super Power" Coast was built for San Pablo, California. It had a 750-gpm Waterous pump and carried 600 gallons of water. Note the equipment cabinets below the floorboards. WAYNE SORENSEN.

Freedom, California, used this 1949 International Model KB-6-Coast with a Waterous pump. The suction hose is extra long and connected to a swivel elbow to allow for quick hookups. WAYNE SORENSEN.

This 1952 International-Coast was one of several built for Los Angeles County. It had a 1,250-gpm Waterous pump and carried 500 gallons of water. The setback front axle improved both turning radius and weight distribution. The R-306 chassis was built in International's Emeryville, California, factory; it included a Hall-Scott engine and was built to carry a fire apparatus body. BOB ALLEN.

Salt Lake County, Utah, used this 1952 International-Coast mounted on an International R-306 chassis, which had a 1,000-gpm Waterous pump. The cab was for cold weather; most apparatus of this era had open cabs. WAYNE SORENSEN.

San Jose, California, had two of these 1954 GMC-Coast pumpers, equipped with Waterous 1,000-gpm pumps. Each pumper carried 1,250 feet of 2½-inch and 200 feet of 1½-inch hose. Both pumpers had been assigned to San Jose by the Office of Civil Defense. WAYNE SORENSEN.

Grants Pass, Oregon, used this 1952 Peterbilt-Coast pumper. It had a 250-gallon booster tank, a 1,250-gpm Waterous pump, and a 480 Hall-Scott motor. COAST APPARATUS, INC.

This 1953 Coast was mounted on an International chassis, and ran as Oakland, California's Engine 7. It had a Hall-Scott motor, a 1,250-gpm Hale double-stage pump, plus a 100-gpm booster pump and a 300-gallon tank. WAYNE SORENSEN.

This Peterbilt-Coast carried a 1,000-gpm Waterous pump and was built for the Burbank Fire District, near San Jose, California. Later, it was sold to nearby Branciforte Fire Protection District. Note the large equipment compartments. WAYNE SORENSEN.

36 • *Motorized Fire Apparatus of the West*

A 1956 International R equipped by Coast with a 250-gpm front-mounted pump and a 400-gallon water tank. Its purchaser was Ben Lomond (California) Fire Department. WAYNE SORENSEN.

San Jose, California, used this 1960 Ford tilt-cab cab-over-engine that Coast equipped with a 750-gpm Waterous pump and a 500-gallon water tank. WAYNE SORENSEN.

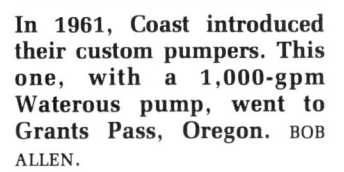

In 1961, Coast introduced their custom pumpers. This one, with a 1,000-gpm Waterous pump, went to Grants Pass, Oregon. BOB ALLEN.

Motorized Fire Apparatus of the West • 37

The Crown Body & Coach plant about 1920. CROWN BODY & COACH.

Crown built the fire fighting body on this mid-1930s Indiana (built by White) for the Los Angeles County Department of Forestry.

In 1947, Crown built the body and front-mounted 500-gpm pump on this International KB-B for use by the Monterey (California) Fire Department. Later, the rig was used by the neighboring Pacific Grove Fire Department. WAYNE SORENSEN.

This 1953 International-Crown 500-gpm pumper, used by Freedom, California, was the prototype for the Crown Firecoach. WAYNE SORENSEN.

Crown Body & Coach Corporation
Los Angeles, California

This firm dates from the turn of the century, when it built carriages and wagons. For the first half of this century, it built a large variety of truck bodies. In the 1930s the firm introduced a school bus with a midship engine. This proved to be popular, and today Crown is best known as a manufacturer of school buses.

Prior to World War II, Crown outfitted commercial chassis with fire apparatus bodies. In 1949, Crown introduced their own custom rig and named it the "Crown Firecoach." These became very popular in the West, particularly in Southern California. Los Angeles County alone has purchased 270 Crowns. (Almost any TV show or movie filmed in Los Angeles showing fire apparatus will have Crown Firecoaches.) The firecoach was relatively standardized for a custom body, and it was designed to stress visibility, maneuverability, ease of operation and ease of access to both the engine (usually a Hall-Scott) and pump (usually a Waterous).

Until recently, Crown still manufactured chassis for pumpers, and entered into several marketing and production agreements with the makers of Pierce apparatus, who are located in Wisconsin. Crown had sold Pierce apparatus on commercial chassis under the name Crown-Pierce.

West Covina, California, purchased the first Crown Firecoach in 1951, a 1,250-gpm pumper, serial number F1001. It had been used for a factory demonstrator from 1951 until 1954. Firecoaches used bolted construction, which made it relatively easy to replace mechancial and body parts. This rig is now owned by Larry Arnold. LARRY ARNOLD.

Motorized Fire Apparatus of the West • 39

This 1953 Crown Firecoach went to American Fork, Utah. It was equipped with a 1,250-gpm Waterous pump, Spicer transmission and a 500-gallon water tank. WAYNE SORENSEN.

Los Angeles Fire Department's High-Pressure Wagon No. 1, a 1955 Crown, equipped with two Byron-Jackson two-stage centrifugal pumps, which could produce 600 psi through one-inch line and fog nozzles. The Siamese 2½-inch inlets on the side deliver water to the turret gun. BOB ALLEN.

This Crown Firecoach was for Azusa, California. It came on Crown's standard 175-inch wheelbase for pumpers. Firecoaches also had adjustable suspension. The curved two-piece windshield and up front position for the driver provided excellent vision. Crown school buses also had the driver sitting forward, and this was widely advertised as a safety factor. CROWN BODY & COACH.

Above: Los Angeles Fire Department Truck No. 38, a 1957 Crown tractor (rebuilt from a high-pressure hose wagon) pulling a 1960 Seagrave 100-foot aerial ladder. BOB ALLEN.

Below: This Crown, with an 85-foot snorkel (built by Pitman), was used by the Orange City (California) Fire Department. It also carried a full complement of ground ladders. The date is about 1960. (Pitman later became Snorkel Fire Equipment Company.) CROWN BODY & COACH.

One of the first Curtis bodies was placed on this 1949 International K, and was for the Mason Valley Fire District at Yerington, Nevada. It had a 500-gpm centrifugal pump. L.N. CURTIS & SONS.

This Curtis body was put on a 1953 Ford F-600 for the Mills (California) Fire District near Sacramento. It had a 500-gpm pump and a 500-gallon water tank. WAYNE SORENSEN.

The Salt Lake City branch of L.N. Curtis & Sons delivered this 1953 GMC to Ephraim, Utah. L.N. CURTIS & SONS.

42 • Motorized Fire Apparatus of the West

Battle Mountain, Nevada, used this 1953 Ford-Curtis 500-gpm pumper with a squirrel-tail-mounted suction hose. This volunteer department carried its turnout gear to the scene of the fire, hanging from the side of the hose rack. WAYNE SORENSEN.

L.N. Curtis & Sons
Oakland, California

In the late 1920s, L.N. "Pop" Curtis was a West Coast salesman for a firm which made mechanical resuscitators. In this capacity, he made sales calls on many fire departments, and soon he began representing other lines of fire service products which would mainly be of interest to small, volunteer departments. In 1934, Pop and his two sons, Bill and Jay, formed a partnership, and the firm grew to the point where it had offices in Oakland, Salt Lake City and Seattle.

After World War II, the firm became the western distributor for Maxim (discussed in Chapter Three). However, the Curtis firm felt there was an additional market for lower-cost apparatus built on commercial chassis. So, in 1948 the firm opened a small assembly shop in Oakland. A few years later an arrangement was made with the George Heiser Body Company of Seattle to assemble apparatus there to Curtis specifications.

The Curtis shop in Oakland was on Oakport Street, and then moved to Peralta Street. In the 1950s the assembly operations were turned over to Earl Sherman & Company, an Oakland truck body builder. Apparatus was placed on many makes of commercial chassis, and Maxim sometimes provided many of the components.

The L.N. Curtis firm continues to this day, and sells equipment to fire departments in nine Western states. It is a dealer for FMC apparatus.

Scotts Valley, California, located in the Santa Cruz Mountains, used this 1957 GMC-Curtis with all-wheel-drive. It had a 750-gpm pump and a 750-gallon water tank. L.N. CURTIS & SONS.

Motorized Fire Apparatus of the West • 43

Above: This 900-gallon tanker is a 1957 International-Curtis with a front-mounted 750-gpm pump. Notice that the bumper is built out in the front to support the pump and pump controls. The rig was for the Aptos (California) Fire Department. WAYNE SORENSEN.

Below: Lakeport County Fire Department at Lakeport, California, used this GMC-Curtis. It had a 750-gpm pump, a 500-gallon water tank, and dual hose reels. WAYNE SORENSEN.

44 • *Motorized Fire Apparatus of the West*

The first Curtis-Heiser truck built was this 1,250-gpm pumper on a Kenworth chassis, delivered to Tumwater, Washington, in 1948. It was powered by a Hall-Scott engine. The rig had a 500-gallon water tank. BILL HATTERSLEY.

Curtis-Heiser
Seattle, Washington

The George Heiser Body Company of Seattle started building fire apparatus on commercial chassis about 1940. They had a close relationship with Kenworth and supplied components for Kenworth bodies and for a limited number of fire trucks Kenworth produced.

In the early 1950s, under a contract arrangement with the L.N. Curtis & Sons firm, of Oakland, California, the Heiser firm built a few apparatus. They also apparently built some apparatus independently of the Curtis agreement.

The firm exists today, although in 1980 they ceased production of fire apparatus.

The George Heiser firm built this truck for Victoria, British Columbia. It's a city service truck on a 1955 Kenworth chassis and has a booster tank and hose reels. On the running board note the protector and windshield for standing fireman. BOB ALLEN.

Motorized Fire Apparatus of the West • 45

Ford did not build truck chassis until 1917. Here is a 1914 Ford auto roadster chassis that F.A.B. Manufacturing Company (now FABCO) used to build a two-tank chemical and hose car for Alameda, California. It also carried a small pump on the side. The roadster trunk was removed and the frame lengthened and straightened. The wheelbase was extended. RAILWAY NEGATIVE EXCHANGE.

San Francisco's Tank Wagon No. 9, was built by FABCO in 1940. It was equipped with a 415-gallon water tank and a Viking 90-gpm rotary pump. The motor was a Hall-Scott 177, six cylinders and rated at 175 hp. BOB ALLEN.

One of the FABCO custom chassis, a 1941 pumper that was delivered to Davis, California. It had a 1,000-gpm pump and a 400-gallon water tank. The side hose compartment was known as a knuckle-buster. In 1955, Van Pelt rebuilt this unit. It still is used in parades. FABCO.

46 • Motorized Fire Apparatus of the West

In 1942, Dixon, California, bought this custom FABCO. It was powered by a Hall-Scott 175-hp engine and had a 750-gpm Waterous pump. WAYNE SORENSEN.

FABCO
Emeryville, California

FABCO started early in this century with its original name being the F.A.B. Manufacturing Company, the three letters coming from the names of the founders: Freitag, Ainsworth and Bean. At first, the firm lengthened Ford Model T and TT chassis for carrying bus and lumber bodies, and eventually they spread into building other types of bodies as well. The firm exists to this day, making four-wheel-drive conversions, agricultural trucks, etc.

In the 1930s, they began mounting fire apparatus bodies onto commercial truck chassis. In 1939, they built three pumpers and two tank wagons on their own custom chassis. These custom rigs went to Davis, Dixon, Emeryville and San Francisco, all in California. During World War II, the firm had a government contract to outfit several hundred Dodge chassis with fire fighting equipment. In addition, they built several types of trailer pumps and small, three-wheel fire apparatus.

After World War II, FABCO sold the fire apparatus portion of its business to two employees, who founded Coast Apparatus, Inc., which was discussed earlier. FABCO later became a division of Kelsey-Hayes.

FABCO built this 1938 Ford for the fire department in Gridley, California. WAYNE SORENSEN.

Motorized Fire Apparatus of the West • 47

Niles (now a part of Fremont), California, used this 1939 Diamond-T-FABCO. (In this era, Diamond-T was possibly the most attractive commercial chassis manufactured in the United States, which may explain one reason it was selected for equipping with fire apparatus.) WAYNE SORENSEN.

This is a 1941 Dodge-FABCO, with a 500-gpm pump, that was used by the Ophir Hills (California) Fire Department. WAYNE SORENSEN.

This 1941 White-FABCO 500-gpm pumper went to Arcade, California. It carried a 750-gallon water tank, large for that time. WAYNE SORENSEN.

Above: **This Chevrolet, from the early 1940s, was outfitted by FABCO for Walnut Grove, California. It had a 500-gpm triple combination pump.** WAYNE SORENSEN.

Below: **1941 Autocar-FABCO, built for Piedmont, California. It is a "quad," meaning that it was equipped to perform four different functions. The lower part of the body contained a water tank and a 250-gpm pump, the middle level was for the ladders and the top of the body was the hose rack. (While well-equipped, the quad's disadvantage was that it could not be in two places at one time; if it were near a hydrant so that it could pump water, then it might be some distance from the fire where its other equipment, say, ladders, might be needed.)** PAUL DARRELL.

Motorized Fire Apparatus of the West • 49

Diamond Springs, California, used this 1944 GMC with a John Bean high-pressure pump and a 500-gallon water tank. Truck chassis built during World War II had no chrome, and resulted in drab fire apparatus. WAYNE SORENSEN.

This 1947 Dodge KW had John Bean high-pressure equipment and a 350-gallon water tank. It was used by the Lake Shore Fire Department, near Clear Lake, California. WAYNE SORENSEN.

This Diamond-T-FMC was used first by Saratoga, California, and then went to Hammet, Idaho. It was equipped with a conventional 500-gpm centrifugal pump, plus a John Bean high-pressure pump. Seats on the front fenders and sweeps under the bumper allowed the truck to attack brush and grass fires while "on the move." GAYLE SORENSEN.

50 • *Motorized Fire Apparatus of the West*

Rio Nido, California's 1948 Chevrolet with a John Bean positive displacement high-pressure pump and a 300-gallon water tank. BOB ALLEN.

FMC (Food Machinery Corporation) San Jose, California

Los Gatos, California, is the birthplace of the Food Machinery Corporation. John Bean, originally from Michigan, moved to Los Gatos from Springfield, Ohio. He purchased an orchard and became very discouraged with the inability of the then-existing pumps to properly spray fungicides in his orchard. He already had some experience with developing pumps because he had invented a continuous flow turbine pump used in conjunction with windmills throughout the Midwest.

Working on the same principle of continuous flow, Bean eventually developed a high-pressure spray pump which was very useful for orchard work. The pump was patented in 1904 under the name Bean Spray "Magic Pump." (In 1902, the FMC firm, which Bean had been instrumental in founding, had been moved from Los Gatos to nearby San Jose, California, on Julian Street.)

The story is told that the effectiveness of the Bean pump for fighting fires was discovered when an orange grower — spraying a flammable oily substance on his citrus trees with a John Bean pump — happened to turn, and was dismayed to see that one of his orchard sheds was on fire. Because he thought the structure was beyond saving, he decided to make it burn faster so that it would be less likely to endanger his grove. So he sprayed oil on the burning structure to speed the burning process, but — to his surprise — the high-pressure spray extinguished the flames. This experience led to the use of high-pressure systems for combating fire.

The firm began outfitting fire apparatus in 1927, using the high-pressure equipment. The high-pressure created a smothering "fog" effect which was useful in combating high-temperature petroleum fires. Also, with high-pressure, a limited water supply could be much more effectively used when fighting a fire. FMC built airport crash trucks for the military during World War II. The firm also built fire apparatus for use by regular fire departments. These rigs had high-pressure or regular pumps (and sometimes both).

In 1973, the firm moved its fire apparatus manufacturing operations from San Jose to Tipton, Indiana. Much later (1978) FMC acquired the Van Pelt operation in Oakdale, California (which is discussed later). FMC exists today, and still manufactures food machinery, fire apparatus and equipment for the military. The Van Pelt plant at Oakdale closed in 1987.

Marin County, just north of San Francisco, used this 1953 GMC with a high pressure pump. WAYNE SORENSEN.

1956 Dodge-FMC was used by the fire department in Kelseyville, California, near Clear Lake. WAYNE SORENSEN.

FMC advertisement from a 1955 trade journal shows a GMC with an FMC volume pump. Ad says the pumpers are "custom-built to western needs." BLACKHAWK AUTOMOBILE COLLECTION.

Cloverdale, California's 1954 GMC-FMC, with both high-pressure equipment and conventional pumps. The majority of FMC apparatus had both volume and high-pressure pumps. High-pressure was used when water was in short supply, or when fighting small fires inside buildings where it was possible to minimize water damage, or for fighting auto fires where fog was more effective than a direct stream of water. WAYNE SORENSEN.

52 • Motorized Fire Apparatus of the West

General, when in St. Louis, outfitted Ford TTs. This is one of their pumpers. CHUCK RHOADS.

General Fire Truck Corporation
St. Louis, Missouri (and, later, Detroit, Michigan)

By virtue of its early years in St. Louis, this firm just barely fits under our "Western" umbrella. It began in 1903 when Harry W. Huthsing founded the National Belting and Hose Company in St. Louis. Soon, its name was changed to General Manufacturing Company and they became distributors for hose and other fire department items. In 1918, they took over the Frank and Saint Gem Fire Extinguisher Company, and they outfitted many Ford TT chassis with fire fighting bodies. These were sold to many small communities in the vicinity of St. Louis.

In 1923, the firm decided to concentrate its energies on production of fire extinguishers and chemical carts, which at that time were little more than truck-mounted soda-acid fire extinguishers. In the late 1920s, the firm collaborated with Pierce-Arrow to build apparatus on the regal Pierce-Arrow "Z" bus chassis. Studebaker bus chassis were also used. (In the late '20s the Studebaker and Pierce-Arrow firms were related themselves.) General also began producing their own "custom" apparatus and named it "General Monarch." Most of the custom apparatus was powered with Hercules engines.

In 1936 the firm moved to Detroit. In 1937 a subsidiary was established in Los Angeles. Then, in 1942 the firm's name was changed to General Detroit, with its Los Angeles subsidiary becoming the General-Pacific Corporation. (The term "General-of-the-Pacific" is often applied to apparatus built at the Los Angeles site.) During 1957 General went out of the fire engine manufacturing business and the firm's name was changed to General Fire Extinguisher Corporation. Several former employees formed their own company to continue the manufacture of apparatus. Their firm is known as F.T.I. (Fire Truck, Incorporated).

General marketed a line of assembled, custom apparatus under the name "General Monarch." Oakland, California, used this 1933 General Monarch chemical car, powered by a 200-hp Hercules engine. In addition to the 50-gallon chemical tanks, a 40-gallon foam tank supplied foam through a single line of one-inch hose. PAUL DARRELL.

Motorized Fire Apparatus of the West • 53

General built this fire engine on a 1938 Packard chassis for use by the Santa Barbara (California) Fire Department. It was powered by a Packard V-12 engine, had a 750-gpm pump and a 200-gallon booster tank. Walt MaCall, the famous fire apparatus historian, aptly referred to this rig as "a true classic among fire engines." It is now in the hands of a private collector. ED GARDINER.

Roswell, New Mexico, bought this 1939 750-gpm pumper from General Detroit. NATIONAL AUTOMOTIVE HISTORY COLLECTION, DETROIT PUBLIC LIBRARY.

General-of-the-Pacific built this rig on a 1940 Kenworth chassis for the Weiser (Idaho) Fire Department. It had a 1,000-gpm pump and a 750-gallon water tank. WAYNE SORENSEN.

54 • Motorized Fire Apparatus of the West

General-of-the-Pacific outfitted this 1946 Dodge series WFA as a 500-gpm pumper for the Benicia (California) Fire Department. The rig is still in use. WAYNE SORENSEN.

This 1949 General-of-the-Pacific was built on an Available chassis and powered by a Waukesha motor. It had a 500-gallon water tank and a 1,000-gpm pump. Its user was Universal City, a large motion-picture studio in the Los Angeles area. WAYNE SORENSEN.

Marysville, Washington, purchased this 1953 Federal General-of-the-Pacific with a 750-gpm pump and a 750-gallon water tank. Note the "CD" emblem on door, which stood for "Civil Defense" and meant the vehicle was part of an emergency vehicle "pool" that could be drawn upon in case of an area-wide emergency. JOHN SORENSEN.

Above: Lake Grove, Oregon, bought this 1955 General-of-the-Pacific, mounted on a Duplex chassis. The chassis had been built in Winona, Minnesota. The truck had a 750-gpm pump and a 750-gallon tank. BILL HATTERSLEY.

Below: This 1912 pumper was built by Gorham for the Oakland (California)Fire Department. It was rated at 1,000-gpm utilizing the discharge gates at the rear. The purchase price was $9,500 and the engine was in service until 1933. The rig had a 14-foot wheelbase and was 23 feet long. It was over seven feet high and weighed 8½ tons. Its top speed was reputed to be 65 mph. This was the prototype for the Seagrave-Gorham WC 144 (water-cooled, 144-hp motor). This centrifugal pumper set pumping engine standards that have endured to this day. OAKLAND FIRE DEPARTMENT.

56 • *Motorized Fire Apparatus of the West*

Pasadena, California, bought this second Gorham pump. PASADENA FIRE DEPARTMENT.

Gorham Engineering Company
Oakland, California

The Gorham Fire Apparatus Company in San Francisco was the West Coast distributor for the Seagrave Company. Gorham Fire was also related to the Gorham Engineering Company in Oakland, California, the manufacturer of large, multi-stage turbine centrifugal pumps.

Seagrave, then of Columbus, Ohio, was also probably involved in the development and design of this apparatus, although we don't know this for certain. The Oakland Fire Department purchased the first such rig in 1912, and it was carried on its records as a Gorham. Pasadena, California, bought the second rig, and also listed it as a Gorham. In a newspaper story appearing in the *Madera* (California) *Daily Mercury* of August 25, 1914, the Pasadena rig was pictured and the story's headline said: "California Built Engine to be Examined by Committee," and continued that the engine was "made in Oakland." Simultaneously, and elsewhere, similar-appearing rigs — with Gorham pumps — were marketed as Seagraves. Now, there is uncertainty as to which firm should get credit for what.

The Gorham pump was different from the rotary gear and piston pumps in general use on fire apparatus. The Gorham pump was a multi-stage centrifugal pump similar to the turbine pumps on steam-driven ocean vessels. The pump could deliver huge quantities of water at very high pressures. Also, it could pump water which contained solids such as pebbles or sand without injury to the pump. (Conventional apparatus pumps had to place strainers on the ends of suction hose.) Hence, this meant the Gorham pump was better adapted to pumping water from ponds or cisterns.

During its brief stint of popularity, 40 Gorham or Seagrave-Gorham pumps were sold, most of them in the West. California cities which purchased the model WC (water-cooled) 96 750-gpm pumps were: Bakersfield, Berkeley, Colton, Los Angeles (two units), Sacramento, San Diego (three), San Jose, South Pasadena, Stockton (two) and Visalia. Outside California, units were sold to Mason City, Iowa; Middletown, New York; Ada, Oklahoma; Wichita Falls, Texas; Seattle, Washington (two); Windsor, Ontario; and Regina, Saskatchewan. WC 144 pumping units, which could pump 1,000 gpm, were delivered to the following California cities: Long Beach, Los Angeles (four units), Napa, Oakland, Pasadena (two) and San Diego (two). Five units also were sold outside California: to Chicago, Illinois; Columbus, Ohio; Tacoma, Washington; and Montreal, Quebec (two). The Gorham centrifugal pump is believed to be the first to be used on motorized apparatus, and this must count as one of the West's most important contributions to the development of fire apparatus in this century.

The Gorham rigs were unique in appearance because nearly all their carrying space was devoted to a huge pump. Both Gorhams and Seagrave-Gorhams are pictured in this chapter. (Other Seagrave rigs are shown in the following chapter.)

This Seagrave-Gorham, serial number 9016, was sold to Sacramento, California, in 1913. It was a Model WC 96, (water-cooled, 96-hp motor). All controls were at the rear. This rig was Sacramento's first motor pumper. Currently, it belongs to the Imperial Palace Hotel in Las Vegas. ED GARDINER.

Just before World War II, Bakersfield, California's 1913 rig, WC 96 (water-cooled, 96-hp motor), serial number 9015, was rebuilt at Hall-Scott plant in Oakland. The motor was replaced with a Hall-Scott motor. ROBERT SAMS.

Colton, California's 1914 Seagrave-Gorham, serial number 10439, going through its acceptance tests at the time of delivery. This rig has a chemical tank on the top of the pump and hoseboxes on either side. DALE MAGEE.

58 • Motorized Fire Apparatus of the West

Berkeley, California's Gorham-equipped Seagrave was a straight pumping engine, carrying no hose. It was a Model WC 96 (water-cooled, 96-hp motor). This picture was taken after the unit was rebuilt in the late 1930s by American Car & Foundry and given a new Hall-Scott engine. PAUL DARRELL.

A rear view of San Jose's (California) 1914 Seagrave-Gorham Model WC 96 (water-cooled, 96-hp motor), serial number 11953. The unit was assigned as San Jose's Engine 1, but proved to be unreliable in the hands of firemen who were accustomed to horses. It required both a driver and engineer to operate. In February 1915, it was banished to another station; in March the San Jose Fire Commission asked Gorham to send a person to once again give instructions as to how to use the machine. SAN JOSE FIRE DEPARTMENT.

In 1938 San Jose's unit, serial number 11953, was rebuilt by Hall-Scott, on a GMC frame, at a cost of $6,300. Very little of the original rig remained except the pump. After the rebuild, it was hard to steer. The rig was kept until 1947. (Because of its hump, San Jose firemen referred to it as the "White Elephant." They may have had additional reasons.) SAN JOSE FIRE DEPARTMENT.

Motorized Fire Apparatus of the West • 59

HALE MOTOR FIRE APPARATUS

Combines Strength, Power, Reliability and Efficiency

Write for prices and details on your requirements before you buy

GEO. C. HALE

OFFICE AND FACTORY:

119-121 W. 14th STREET   KANSAS CITY, MO.

Above: This is believed to be the first motorized apparatus built by George C. Hale, a combination chemical and hose car for Tulsa, Oklahoma. FIREMEN'S HERALD.

Below: This 1935 Dodge hose wagon was outfitted by Hedberg Manufacturing of San Jose for the Burbank Fire District in Santa Clara County, California (now the Central Fire Protection District of Santa Clara County). The rig still exists. WAYNE SORENSEN.

60 • *Motorized Fire Apparatus of the West*

Hedberg Manufacturing built this 500-gpm rescue squad on a 1938 Mack chassis for the San Jose Fire Department. This was the first radio-equipped fire apparatus in San Jose. DICK ADELMAN.

Grey Fire Apparatus
Lewiston, Idaho

We know very little about this firm. It also builds road machinery.

The George C. Hale Company
Kansas City, Missouri

George C. Hale was a well-known builder of horse-drawn fire apparatus in the Kansas City, Missouri, area. At one time he had been the city's fire chief, and, before the turn of this century, he patented a telescopic mast elevated by hydraulic pressure of water or soda-acid chemical reaction. The tower was pulled by a team of three horses, and the direction of stream flow could be controlled by one man on the ground.

One of the first pieces of motorized apparatus Hale built was for Tulsa, Oklahoma, in 1913. Eventually, this firm became part of the LaFrance Fire Engine Company.

There is also a Hale Fire Pump Company in Conshohocken, Pennsylvania (mentioned in the following chapter), although we do not know whether the two firms are related.

Hedberg Manufacturing Company
San Jose, California

Founded by J.N. Hedberg, this firm's main claim to fame in the fire apparatus field was its "long-roll" coaster siren. Hedberg also operated a private fire fighting service which was used by dwellers in unincorporated areas of Santa Clara County. For a while the firm also outfitted fire apparatus for use in their local area. The firm is in business today and continues to build sirens.

A chemical-hose combination built by Howard-Cooper on a Ford TT chassis for Milwaukie, Oregon. Ford TT chassis were for truck bodies, they were longer and had lower gearing than did Ford T auto chassis. HOWARD-COOPER.

Motorized Fire Apparatus of the West • 61

Howard-Cooper was a Gramm-Bernstein dealer and used their chassis to build this rig, which came with both chemical tanks and a 400-gpm Northern pump. Gramm-Bernsteins were manufactured in Lima, Ohio. HOWARD-COOPER.

The Glenns Ferry (Idaho) Fire Department decided they needed to carry more water on this Howard-Cooper rig, so they added a 100-gallon booster tank, made from oil drums, on the top of the hose compartment, and they also had to add dual rear wheels to handle the additional weight. There is an oval Howard-Cooper emblem riveted to the Gramm-Bernstein radiator casting containing the names of three cities: Portland, Seattle and Spokane. WAYNE SORENSEN.

Renton, Washington, received this 1927 Howard-Cooper Special, which came with a 450-gpm pump. Note the large water tank. The unit has been restored by the Renton Fire Department. HOWARD-COOPER.

62 • Motorized Fire Apparatus of the West

This 1927 Howard-Cooper is mounted on the Gramm-Kincaid chassis. It is a 450-gpm pumper and was used by the Shelton (Oregon) Fire Department. (Gramm-Kincaid trucks were assembled in 1925 and 1926 in Lima, Ohio.) HOWARD-COOPER.

Howard-Cooper
Seattle, Washington

This firm was founded in 1912 by George Howard and D.I. Cooper to build road-paving equipment. In the 1920s, they began building fire apparatus, often using Gramm-Bernstein chassis. At one time, they were building as many as 40 to 50 units a year. They stopped building apparatus in 1957, according to a present-day company spokesman, because: "We could not compete with the small custom garages."

In Renton, Washington, a Howard-Cooper rig has been restored, and its oval radiator emblem lists these cities: Seattle, Spokane and Portland, Oregon. The firm had offices in Seattle and Portland and, at various times, served as the dealer/distributor for Gramm-Bernstein trucks and fire apparatus, and for Seagrave.

The firm exists today, and is headquartered in Portland. It sells logging, mining and construction equipment.

The Howard-Cooper emblem is on the radiator of this 1928 Reo chemical and hose car which was sold to Ashland, Oregon. HOWARD-COOPER.

Motorized Fire Apparatus of the West • 63

Winlock, Washington, purchased this 1931 Chevrolet M series, which Howard-Cooper outfitted with a front-mounted 350-gpm pump, hose reel and water tank. HOWARD-COOPER.

This 1936 Ford V-8 was outfitted as a pumper by Howard-Cooper for St. Anthony Falls, Idaho. It carried a Seagrave 500-gpm pump and an 80-gallon booster tank. HOWARD-COOPER.

This tank wagon was mounted on a 1941 International Model K chassis by Howard-Cooper. It was equipped with a 150-gpm Oberdorfer pump and its user was Clackamas County, Washington. HOWARD-COOPER.

Above: This 1941 Diamond-T-Howard-Cooper, with a 500-gpm pump, was for Bethel, Washington. In 1941, the Chicago-based Diamond-T Company startled the truck industry by announcing that they were guaranteeing their trucks for an entire year or for 100,000 miles. HOWARD-COOPER.

Below: Howard-Cooper put a 500-gpm pump on its 1951 Diamond-T for the King County (Washington) Fire Department. HOWARD-COOPER.

Motorized Fire Apparatus of the West • 65

NORTHERN STANDARD COMBINATION CHEMICAL AND HOSE CAR

In buying apparatus of this type, you, Mr. Fire Chief and Mr. Committeeman, will take into consideration

QUALITY——RELIABILITY——PRICE

THE NORTHERN CAR

embodies every feature that is destined to reflect credit to your judgment in its installation

MR. FIRE CHIEF
MR. COMMITTEEMAN
{ Your personal pride prompts you to **investigate**, that you may not be criticized by your fellow citizens.

EXTREME ORIGINAL INVESTMENT CAUSES COMMENT
LARGE UPKEEP EXPENSE MEANS CENSURE
UNRELIABILITY BRINGS JUST CRITICISMS

The Northern Car is designed and built to eliminate all possibility of such after-comment.

How can you be sure you are making the best possible deal for your city unless you **investigate**.

You cannot afford to purchase a similar apparatus without investigating "THE NORTHERN."

Let us explain the many points of merit in our car and impart to you our confidence in its ability to serve your city and fire department's best interests more capably than any other.

You will be interested in learning more about this car.

Write us for detailed information and complete specifications.

NORTHERN FIRE APPARATUS CO.
MINNEAPOLIS

An advertisement from the *Firemen's Herald* from about 1912 for Northern Fire Apparatus Company.

In 1922, a Luverne salesman made it far enough west to compete with both American-LaFrance and White sales representatives in making a sales presentation to Carmel, California, city officials. He did the best job, and here's the 1923 Model 6 Luverne pumper he sold to this picturesque community. CARMEL FIRE DEPARTMENT.

Luverne Fire Equipment Company
Luverne, Minnesota

This firm began in 1906, evolving out of a carriage-building operation of two brothers, Al and Ed Leicher, and the new firm built both autos and trucks. In 1913 they built a motorized pumper for the city of Luverne, Minnesota. They stopped building autos in 1917, trucks in 1923. However, they continued to outfit fire fighting apparatus, a function they still perform.

This 1924 Luverne pumper has been restored by the Nobles County Historical Society of Worthington, Minnesota. NOBLES COUNTY HISTORICAL SOCIETY.

Northern Fire Apparatus Company built this double-tanked chemical and hose car for St. Anthony, Idaho. It's on a 1922 Dodge chassis. Note the Budd-Michelin steel disc wheels. FIRE ENGINEERING.

American Fork, Utah, purchased this 1924 International-Northern 250-gpm pumper. FIRE AND WATER ENG.

68 • *Motorized Fire Apparatus of the West*

Petaluma, California, bought this 1912 Nott "Universal" four-cylinder 500-gpm pumper. FREEMAN PHOTO.

Northern Fire Apparatus Company
Minneapolis, Minnesota

This company was active late in the last century and early in this century, and was known best for the steel and copper chemical tanks it made for the various builders of apparatus. Their tanks rotated, and on each rotation another small batch of acid would be added to the soda mixture. At the time of World War I, the firm was also offering its bodies on several makes of commercial chassis, including Ford, FWD, Graham Brothers, International and Packard.

In addition, they built 500-gpm rotary pumps, which were used by both themselves and other well-known chassis outfitters.

Nott Fire Engine Company
Minneapolis, Minnesota

The Nott firm began building automotive pumping engines about 1912. The firm had developed a piston pump suitable for gasoline engine drive. In 1912, Victoria, British Columbia, purchased a Nott which would have been considered as "streamlined" in its day. Nott also built smaller "rotary-roller"dual- or single-action pumps of 600-gpm and 800-gpm capacities.

In 1912, Petaluma, California, also purchased a Nott. This was a "Universal" model which originally had been displayed at fire chiefs' conventions in

Oakland, California, purchased this 1912 Nott 500-gpm pumper. The rig was loaned to the San Francisco Fire Department and used in a timed contest with horse-drawn equipment. The Nott, with a time of three minutes, 40 seconds, won. The horses took five minutes, six seconds. OAKLAND FIRE DEPARTMENT.

Motorized Fire Apparatus of the West • 69

DO YOUR OWN THINKING!

"Universal" 4-Cylinder Motor Pumping Engine.

PETALUMA FIRE DEPARTMENT.

Petaluma, Cal., Oct. 1, 1913.

The Woodhouse Mfg. Co., New York, N. Y.

Gentlemen:—In reply to yours of September 19th, beg to state that I voice the sentiment of the entire Board of Fire Commissioners, City Council and of every local Fireman, when I state that the Nott pumping engine (auto type) is a wonderful machine. Ours is giving the very best of satisfaction and is always ready and "on the job." It has been in commission for a year and has done splendid work despite the fact that it was at first in charge of inexperienced hands.

The engine has done more than the builders realized it could do in tests made by us under unusual conditions. It is holding up perfectly and with the Nott we feel perfectly safe. We do not think that you have on your list any machine that will eclipse the Nott and our whole community are boosters for the engine.

The builders have treated us royally, have made good every promise and in addition have given us several hundred dollars' worth in fittings that we have never asked for or paid for. It is a pleasure to do business with such people, and we feel sure that if you take the agency, you will never have cause to regret such action.

We are very proud of the engine and take great pride in showing it to the numerous visitors, many of whom come here expressly to inspect it. I have never heard a word against it but have heard much praise, and mechanics pronounce it a splendid piece of workmanship. We would not part with it under any conditions and I am truthful in stating that we deem it a most valuable asset.

Respectfully,

President, Board of Fire Commissioners,
H. S. McCARGAR.

R. S. ADAMS,
    Chief Engineer.

(Original copy on file in Minneapolis office.)

THE NOTT FIRE ENGINE CO.
MINNEAPOLIS, MINN.

E. A. WILKINSON
General Manager

D. A. WOODHOUSE, General Eastern Representative,
50 West Broadway, New York.

The *Firemen's Herald* ran this Nott advertisement, which contained a testimonial concerning the Petaluma, California, rig.

The Pacific Fire Extinguisher Company mounted the hose body and booster equipment on this 1930 LR Chevrolet for Livermore, California. PAUL DARRELL.

both Denver and Los Angeles. The unit had four cylinders, and, according to news reports, was welcomed to Petaluma with a giant parade. In a demonstration of its pumping abilities, the pumper managed to create so much pressure that it burst the Petaluma Fire Department's old hose lines, dampening many of the onlooking dignitaries. New hose had to be purchased before the Underwriters' acceptance tests could be completed. Petaluma found that the rig was difficult to steer and the pump was complicated to operate. Also, because of its weight, it broke through the fire hall's floor. However, it gave good service, pumping for more than three hours during a fire in 1914. Petaluma kept their Nott until 1942, when it was scrapped to support the World War II effort.

Pacific Fire Extinguisher Company
Los Angeles, California

This was the Los Angeles and San Francisco sales agent for Ahrens-Fox (Cincinnati), maintaining offices in both cities. The firm also had a small machine shop and they outfitted a few apparatus on commercial chassis. They dropped out of business about 1960.

Watsonville, California, used this 1935 Ford, which had been outfitted by Pacific Fire Extinguisher Company. It had a 500-gpm pump. PAUL DARRELL.

Long Beach, California, ran this 1912 Robinson "Jumbo" pump and hose wagon, which could pump 1,000-gpm. (Robinson used colorful names for its apparatus models: "Jumbo," "Whale," "Invincible" and "Vulcan," to name a few.) ROBINSON FIRE APPARATUS MANUFACTURING COMPANY.

Globe, Arizona, operated this 1914 Robinson pumper with the squirrel-tail suction hose wrapped around the front of the engine. Note the big steamer-type air chamber and bell on the dash. The photo also shows the triplex pump drive chain, a crude power take-off to the pump. ROBERT SAMS.

72 • Motorized Fire Apparatus of the West

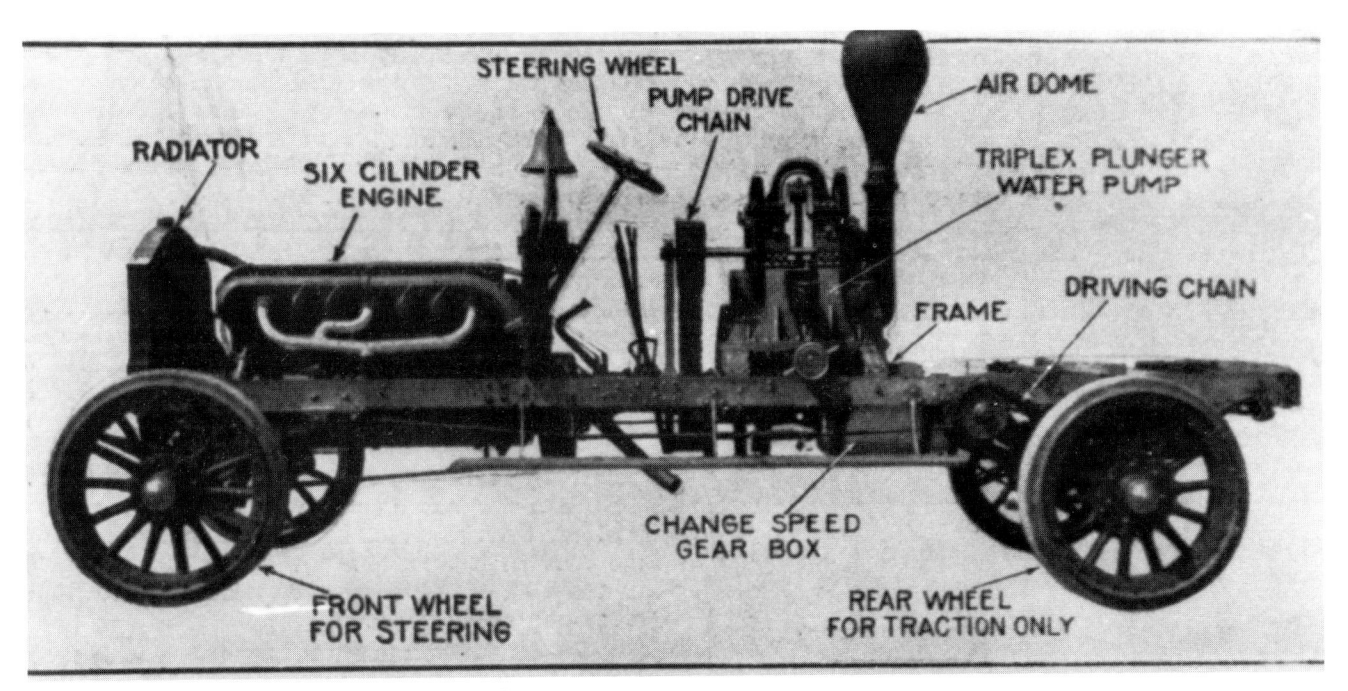

This advertisement shows a "stripped" Robinson.

Robinson Fire Apparatus Manufacturing Company
St. Louis, Missouri

Founded just after the Civil War, this firm is considered to be one of the real pioneers of American fire apparatus manufacturing. It had been formed in 1871 out of two other firms: the P.J. Cooney Company and the Stemple Fire Extinguisher Company. It built a full line of horse-drawn apparatus: chemical wagons, combination hose and chemical rigs and ladder trucks. Their products served fire departments throughout the nation and, for a time, were very popular in the West.

In 1907, they built their first motorized rig, a buckboard-style combination hose and chemical wagon. Soon they were building more motorized units, utilizing large passenger car chassis built by the Chadwich Engineering Works of Philadelphia. Power came from Buffalo marine-type motors, and Robinson provided its own pumps. In 1916, the firm announced that it would also produce large commercial motor trucks, although little can be found concerning them. Apparently, the firm manufactured fire apparatus up until 1932, although hardly anything is heard of its products during its latter years.

(In 1984, considerable attention was given to a 1911 Robinson pumper restored for the Staunton Fire Department in Virginia. The rig had a 1,245-cubic-inch Buffalo marine engine which burned 95 gallons of gasoline per hour, when under full load.)

Roney
Portland, Oregon

This was an Oregon-based sales agent for Howard-Cooper, which outfitted some apparatus on commercial chassis. Equipment was built in the shops of Westland Trailers, of Portland.

Roney outfitted this 1957 Diamond-T with 500-gpm pump and booster tank for Summit, Washington. RONEY FIRE EQUIPMENT.

Above: Redwood City, California, purchased this 1915 Schneer. It was a two-tank chemical and hose wagon. Today, the rear wheels remain; they're in the Redwood City Fire Department's Museum. REDWOOD CITY FIRE DEPARTMENT.

Below: This 1915 Schneer — powered by an 80-hp six-cylinder Wisconsin engine — was originally intended for the San Francisco Fire Department, but rejected because its pumping capacity was inadequate. St. Helena, California, then bought the rig and used it for many years. It has been restored by the St. Helena Fire Department. WAYNE SORENSEN.

74 • Motorized Fire Apparatus of the West

This, the first Van Pelt, was built for Oakdale, California, on a 1925 Graham Brothers chassis. P.E. Van Pelt is at the wheel. The riveted water tank is square. On the front was a red warning flag, the only warning device. There were no red lights or sirens.
ROBERT SAMS.

J.J. Schneer Company
San Francisco, California

The J.J. Schneer Company built a small number of passenger cars and a few pieces of fire apparatus. The fire apparatus used Wisconsin or Continental six-cylinder engines. In 1915, San Francisco purchased a Schneer chemical and hose car, as did nearby Redwood City.

San Francisco also purchased three Schneer tractors in 1915-16 to pull ladder trucks. The tractors were linked to a 1903 Kinney and to a 1911 Larkin (both built in San Francisco) and to a 1903 American-LaFrance.

St. Helena, California, took delivery of a 500-gpm pumper originally intended for San Francisco, but which San Francisco had rejected because the pump was of insufficient capacity. The St. Helena Fire Department kept and has restored the rig; it is currently on display in their fire station. This is the only known surviving Schneer fire truck.

P.E. Van Pelt
Oakdale, California

Van Pelt has been a respected and well-known name in West Coast apparatus since 1925. P.E. Van Pelt, an auto dealer in Oakdale, California, was approached by a local fire chief, O.C. Bailey, and others with a request that he help them in constructing a fire truck. A part of his dealership's garage was set aside for the project and work began. A Graham Brothers (similar to Dodge) chassis was used to carry a water tank, and a smaller separate pump was placed on the vehicle's left running board. The truck was also outfitted with a siren and a red flag to warn of its approach. It was used for fighting fires in rural areas.

This is the third Van Pelt truck, and the customer was Elk Grove, California. It's on a 1925 Graham Brothers chassis. Note the round tank. It holds 500 gallons, and was "hot riveted." The rig was later sold to Oakdale, California, and has been restored.
WAYNE SORENSEN.

Motorized Fire Apparatus of the West • 75

Danville, California, purchased this 1931 Autocar-Van Pelt with a 500-gpm pump and 500-gallon water tank. WAYNE SORENSEN.

Van Pelt used a 1930 Reo chassis to build this 500-gpm front-mount pump for Modesto, California. Note the helmet rack on the side of the rig. BOB ALLEN.

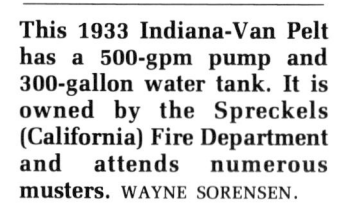

This 1933 Indiana-Van Pelt has a 500-gpm pump and 300-gallon water tank. It is owned by the Spreckels (California) Fire Department and attends numerous musters. WAYNE SORENSEN.

76 • *Motorized Fire Apparatus of the West*

Walnut Creek, California, used this 1935 White-Van Pelt with a 350-gpm Berkeley front-mount pump and 300-gallon water tank. Note the hose reels out toward the rear. PAUL DARRELL.

On the day of the truck's completion, there was a fire in Oakdale, California, and the rig responded with P.E. Van Pelt himself at the wheel. The piece performed well, and Van Pelt decided to build a second truck. This was soon sold to a group of farmers in Waterford, California. A third vehicle was sold to Elk Grove, California, and so on. Soon, Van Pelt was devoting all his energies to producing and selling fire engines.

Part of his success dealt with his ability to hot rivet the steel tanks carried on trucks. Up until that time, riveted tanks on fire trucks (and in other liquid-carrying applications) were generally unsatisfactory because the vibrations from the roads would shake some rivets loose. This was also a time when many apparatus outfitters would not incorporate water tanks into their designs. It turned out that much of Van Pelt's market would be rural fire departments which wanted to carry water tanks. Small Berkeley pumps, manufactured by the Berkeley (California) Pump Company, and ranging in capacity from 250 to 400 gpm,

The Magalia Volunteer Fire Department in Butte County, California, bought this 1936 Dodge-Van Pelt. It has a high-pressure pump. WAYNE SORENSEN.

Motorized Fire Apparatus of the West • 77

Stanford University, Palo Alto, California, once had its own fire department, although now they're served by the Palo Alto Fire Department. Here's a 1936 Ford-Van Pelt quad Stanford purchased. It had a 500-gpm Darley pump. PAUL DARRELL.

Fairfax, California, was the user of this 1937 Ford-Van Pelt. It came with a 150-gallon water tank and a 500-gpm Hale pump. It has been restored. PAUL DARRELL.

Union City, California, still owns this distinctive 1938 Dodge-Van Pelt, which came with a 500-gpm Hale pump and a 200-gallon water tank. WAYNE SORENSEN.

78 • *Motorized Fire Apparatus of the West*

Van Pelt built this 500-gpm pumper for San Juan Bautista, California, on a 1939 International D series chassis. The pump is a Barton. The long equipment box has a screen, rather than being covered. WAYNE SORENSEN.

would be mounted at a rig's front. Van Pelt can claim some credit in the move which shifted booster equipment from chemical tanks to water tanks with small pumps.

Van Pelt also built small pumpers on commercial chassis using Hale midship pumps. During World War II, he manufactured several large orders for the military, mostly on White chassis. After the war, Van Pelt's son-in-law, Nip Dallas, assumed many of the responsibilities for running the firm. Gradually, they expanded their market to include large city — as well as rural—fire departments. In 1952, the firm delivered an order of 13 1,000-gpm pumpers on GMC chassis to the California State Office of Civil Defense. In the 1950s, the firm was using its own plane to fly replacement parts. In 1960, they built their "custom" rig.

P.E. Van Pelt died in 1963. In 1978, the firm was acquired by FMC. In mid-1986 it was announced that the firm's production would be shifted to Tipton, Indiana, and that the Oakdale facility would remain in use for repair and maintenance work. FMC closed the Van Pelt plant July 2, 1987. Former Van Pelt employees are operating the old Van Pelt operation in Oakdale. They use the name Hightech and they are doing apparatus repair work and building some new apparatus.

Richmond, California, bought this 1940 Ford-Van Pelt city service truck. Note its heavy load of ladders, lights, and rescue and first aid equipment. Later, it was sold to a small-town department. WAYNE SORENSEN.

Motorized Fire Apparatus of the West • 79

Soquel, California, purchased this 1947 Dodge-Van Pelt, which had a 350-gpm front-mount pump and a 350-gallon water tank. This picture was taken at a muster when the retired rig was owned by Chris Cavette, who was, and is, newsletter editor for the California Chapter of the Society for the Preservation and Appreciation of Antique Motor Fire Apparatus in America. Posters encouraging membership are shown in front of the unit. DON WOOD.

Van Pelt used a 1947 K-series International to mount a 500-gpm Hale pump and a 300-gallon water tank for Carmel Highlands, California. WAYNE SORENSEN.

This is a 1948 Kenworth-Van Pelt, with a 1,000-gpm pump, used by the Modesto (California) Fire Department. Deck pipe on the top of the hose body provided both volume and range for large fires, such as at lumber yards. WAYNE SORENSEN.

80 • Motorized Fire Apparatus of the West

Ben Lomond, California, used this 1948 White COE (cab-over-engine) 1,000-gallon tanker, which was outfitted by Van Pelt. A short wheel base was needed for the area's narrow mountain roads. WAYNE SORENSEN.

This short wheelbased 1948 Dodge-Van Pelt was used by the Aptos (California) Fire District. It came with a 650-gpm pump and a 500-gallon tank. WAYNE SORENSEN.

The Galt Fire District in Galt, California, operated this 1948 Chevrolet-Van Pelt, equipped with a 350-gpm front-mount pump. This equipment was used for grass and brush fires. Note the seat on the right front fender that allowed a fireman to sit there and train water from a small hose on the fire as the truck moved along. The long equipment box on the left running board is common to most Van Pelt apparatus. WAYNE SORENSEN.

Motorized Fire Apparatus of the West • 81

The Central Fire District of Santa Clara County, California, still uses this 1949 Kenworth-Van Pelt as a parade piece. It has a Hall-Scott motor, a 500-gallon water tank and a 1,250-gpm Hale pump, and was in service until 1977. WAYNE SORENSEN.

This 1950 White, with a power-tilt cab, was used by Van Pelt to build a 600-gpm pumper for Castroville, California. The unit carried a 650-gallon water tank. WAYNE SORENSEN.

Kern County, California, purchased two of these Peterbilt-Van Pelts in 1949. They have Hall-Scott 470 power, 750-gpm Hale pumps and 500-gallon water tanks. The chassis numbers were M-506 and M-507. WAYNE SORENSEN.

82 • Motorized Fire Apparatus of the West

Modesto, California's Engine 5 was an International-Van Pelt with a Hall-Scott motor and a 1,250-gpm Hale pump. It's on an International R-306 fire apparatus chassis. Notice the semi-recessed installation for twin unit reels for the booster hose. The reels can be rewound electrically. WAYNE SORENSEN.

San Rafael, California, uses this 1958 Diamond-T-Van Pelt, which has a 500-gallon tank and a 1,250-gpm Hale pump. The deck pipe is controlled by a worm-and-gear device. WAYNE SORENSEN.

This is a 1961 Oshkosh four-wheel-drive chassis, equipped by Van Pelt with a 1,250-gpm Hale pump and 500-gallon tank for the Palo Alto (California) Fire Department. More recently, the rig was converted to a tanker. WAYNE SORENSEN.

Motorized Fire Apparatus of the West • 83

Mountain View, California, bought this 1960 Van Pelt "custom," which had a Hall-Scott motor, 1,250-gpm Hale pump and carried 500 gallons of water. This was Van Pelt's first custom rig. It is now in reserve status. WAYNE SORENSEN.

Going a year beyond the time span indicated in our book's title, we include this 1961 International-Van Pelt, with a 50-foot Menco aerial ladder, built for Crockett, California. BOB ALLEN.

Continuing again beyond the time span of this book, this is a 1962 Ford C-Van Pelt with an 85-foot Hi-Ranger mobile aerial, used in Vernon, California. WAYNE SORENSEN.

84 • *Motorized Fire Apparatus of the West*

Vehicle historians must be careful using superlatives. However, in terms of the development of motorized fire apparatus, this is probably the single most significant vehicle pictured in this book—it was the first to use the same engine for both propulsion and pumping. It is a 1906 Waterous, used by the Alameda (California) Fire Department. WAYNE SORENSEN.

Waterous Company
St. Paul, Minnesota

The Waterous Engine Works Company was founded in Brantford, Ontario, in 1844, and it manufactured machinery, sawmill equipment and steam fire engines. In 1881, the firm's founder, C.H. Waterous, sent his twin sons west to Winnipeg, Manitoba, to establish a branch. This soon outgrew the home office. Then, a private land developer in South St. Paul, Minnesota, offered the Waterous brothers free land if they would locate an operation there. They agreed and a plant opened in 1887. The company soon was manufacturing a wide variety of horse-drawn apparatus. Eventually, they became independent from the Canadian firm.

The Waterous firm (for a long time known as the Waterous Fire Engine Works) deserves much of the credit for introducing the gasoline engine to fire fighting use. Even before the turn of the century, they had marketed a gasoline-powered, horse-drawn pump.

A recent company brochure described this by saying:

"This was long before the days of spark plugs, and the ignition system for the single cylinder consisted of a hollow platinum tube screwed into the combustion chamber. The tube was heated with a blowtorch, and as the compression forced vaporized gas into the tube, the charge exploded, and the heavy flywheel carried the piston through the exhaust, intake and compression strokes."

In 1906, the firm delivered to Wayne, Pennsylvania, what is generally believed to be the first motorized fire apparatus in the United States. The rig had two engines, one for pumping and one for propelling the vehicle. At that time, the two engines were needed because there was too wide a difference in the characteristics of an engine that could power a pump and an engine that could power a truck.

Waterous was successful in overcoming these differences, however. And in 1907 it delivered to Alameda, California, the first fire engine with a pump driven by the same engine as moved the vehicle. Walter T. Steinmetz, Alameda's fire chief, wrote a testimonial in 1912 about this engine, citing its low upkeep cost, and adding: "The Waterous Gasoline Pumping Engine has given the City of Alameda perfect service, never failing to respond to an alarm of fire, or pumping at the same, during the four years it has been in our service."

Between 1907 and 1918, Waterous built its own fire apparatus; and from 1918 until 1929 it mounted its pumps and bodies onto commercial chassis. After 1929, it specialized in the manufacture of fire pumps and became a large supplier of pumps used by various apparatus outfitters. The firm exists today.

Shown is the view from the left rear of Alameda's 1906 Waterous. It could pump 600 gpm, but carried no hose. WAYNE SORENSEN COLLECTION.

Seattle, Washington, bought this 1911 Waterous Type 15 motor-driven hose wagon. It could carry over 1,200 feet of 2½-inch hose. SEATTLE FIRE DEPARTMENT.

Wentworth and Irwin built four of these 1938 Fageol 1,000-gpm pumpers for the Portland (Oregon) Fire Department. They were powered by Hall-Scott 177 engines. JIM T. BOYD.

86 • Motorized Fire Apparatus of the West

Portland, Oregon, firemen referred to this enclosed city service truck as a "bread wagon" because of its appearance. Wentworth and Irwin built this on a 1939 Kenworth chassis. BOB ALLEN.

Webb Motor Fire Apparatus Company
St. Louis, Missouri

The Webb Company is discussed in the following chapter. However, for a short time — 1910-1912 — it was headquartered in St. Louis, although we don't know whether any apparatus was manufactured there.

Wentworth and Irwin
Portland, Oregon

Founded in 1903, this firm was a body shop and GMC dealership located in Portland, Oregon, which outfitted some apparatus for use in the vicinity. They also built logging trailers, refrigerated and dry bodies and bus bodies, all sold under the "Wentwin" name. The firm still exists, although currently does not build fire apparatus.

Wentworth and Irwin constructed this city service truck for Portland, Oregon, on a 1954 GMC chassis. BRUCE HOLLINSHEAD.

Motorized Fire Apparatus of the West • 87

Above: Western States Fire Apparatus outfitted this 1956 GMC for Gold Beach, Oregon. WESTERN STATES FIRE APPARATUS, INC.
Below: The Benton County Fire District in Washington state bought this Ford equipped by Western. It has a front-mount pump and is painted a light color. WESTERN STATES FIRE APPARATUS, INC.

This is a Yankee airport crash truck, mounted on a wartime GMC chassis. A foam nozzle is above the cab. MOTOR VEHICLE MANUFACTURERS ASSOCIATION.

Western States Fire Apparatus, Inc.
Cornelius, Oregon

Founded in 1946 by Gloyd Hall as Hall Industries, it took its present name in 1955. This firm both builds custom apparatus and outfits commercial chassis. It operates today.

Yankee
Los Angeles, California

Founded in about 1930 by three brothers, Jim, Ray and Sam Yankee, this company built both fire engines and airport crash trucks. At various times it had arrangements to use Seagrave and Walter chassis. Today, the firm is named Fire-X, and is best known as a builder of airport crash trucks.

This 1952 GMC series 280 was a Yankee foam truck built for the use on the San Francisco-Oakland Bay Bridge. It carried 500 gallons of water, plus foam equipment. BOB ALLEN.

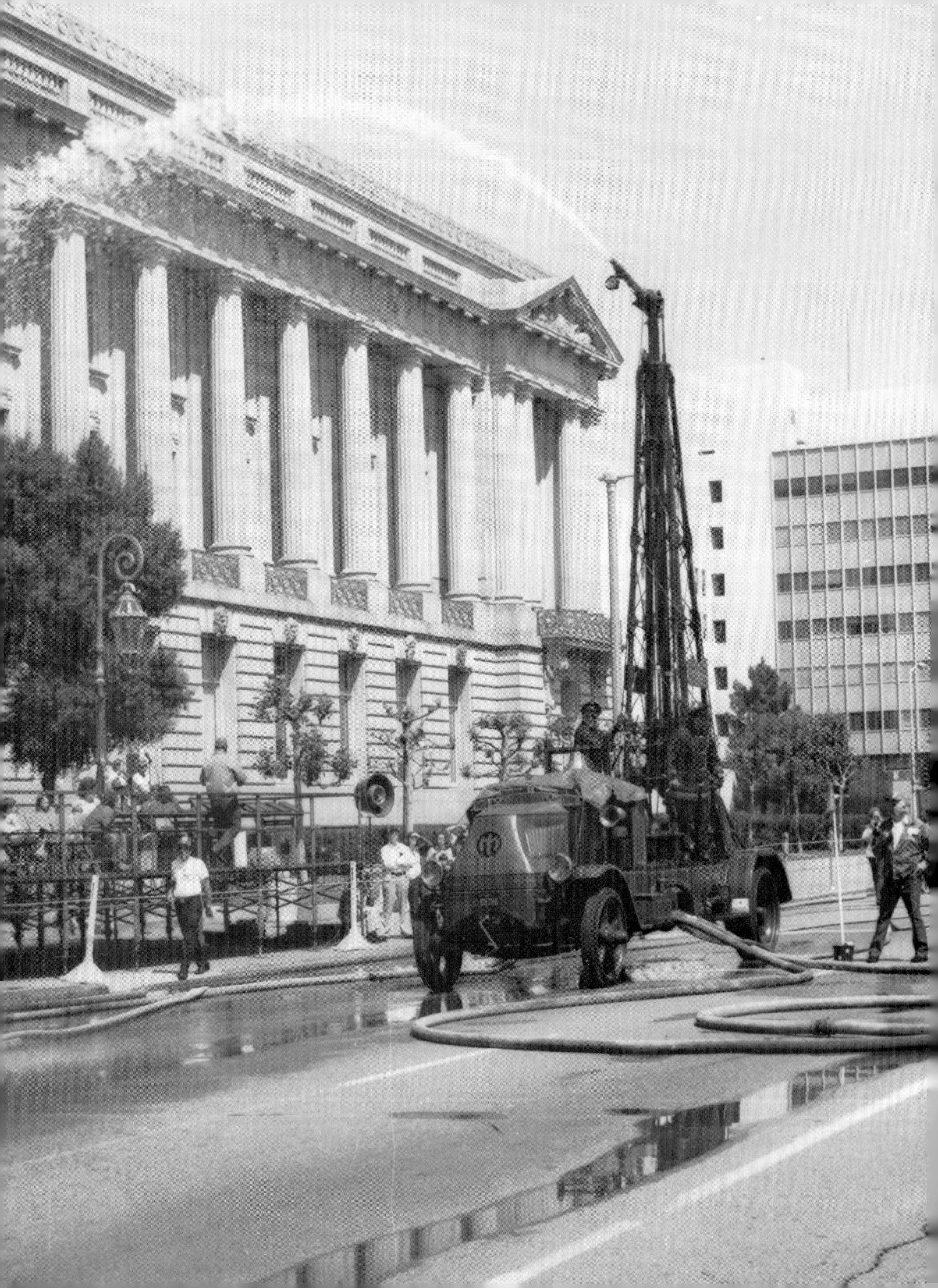

3

The "Eastern" or "National" Manufacturers

This chapter examines those apparatus manufacturers located east of the Mississippi River who were sufficiently prominent that their products were sold in the Far West. For the most part, that means only those Eastern manufacturers whose products could be found west of the Rocky Mountains. We think that that fact alone qualifies them as being somewhat national in scope. Most of the manufacturers described in this chapter made "custom" apparatus. Both that and their applications on commercial chassis are pictured.

Ahrens-Fox
Cincinnati, Ohio

Ahrens-Fox pumpers, with their highly polished globe in front, were probably the most distinctive of all American fire apparatus. The company was founded in 1868 by John P. Ahrens, whose family built steam pumpers, and Charles H. Fox, assistant fire chief of the Cincinnati Fire Department. In 1911, the firm built its first motorized pumper, which had a double-dome piston pump and was powered by a Herschall-Spillman engine. In 1915, Ahrens-Fox converted to a single-dome pump (which continued in production until 1952!).

A line of rotary pumps was developed in 1925 to meet competition. In the '30s, a smaller and less expensive line appeared built on Schacht chassis, and a few more common makes of commercial chassis were also outfitted by Ahrens-Fox. The firm managed to survive the Depression, and during World War II ceased building new apparatus, but continued to service existing models. Eventually, the famed piston pump was too expensive to build and could not compete cost-wise with competitors' centrifugals. Hence, Ahrens-Fox introduced its own centrifugal pump. However, the firm's fortunes ebbed, and fire apparatus production ended in 1957, although the firm had become associated with Mack and there were some "carryovers" into building Mack apparatus.

Right: While Ahrens-Fox is most famous for its piston pumper, the company also produced specialized apparatus. This two-tank chemical unit, Model G (serial number 426), was built for Sacramento, California, in 1914. Each tank held 80 gallons. ED GARDINER.

Preceding Page: The San Francisco Fire Department's water tower shown in a fire fighting stance at a 1976 muster. WAYNE SORENSEN.

The first motorized pumper in San Francisco's fire department was this 1914 Ahrens-Fox "Continental" Model B. The 750-gpm double-domed Continental piston pumper had no hose body; the crew sat inside for greater safety. There were compartments below the seats for tools and other gear. Pneumatic tires were added in 1931. On the right is a 1913 American-LaFrance combination hose and chemical wagon that ran with the Ahrens-Fox. Both were attached to Engine Company 10. Note wooden hose body. BOB ALLEN COLLECTION.

Sacramento, California, used this 1914 Ahrens-Fox Model F, serial number 555, tractor to pull a horse-drawn Seagrave 65-foot aerial ladder. It was powered by a Moore motor. PAUL DARRELL.

Ahrens-Fox featured an Oakland, California, rig in this advertisement. The unit was a 1916 Model K-3 750-gpm piston pumper, which also carried a 40-gallon chemical tank.

92 • *Motorized Fire Apparatus of the West*

One of San Francisco Fire Department's two 1938 Ahrens-Fox Model H-85 aerials. The ladder trailers had electrically-powered hydraulic systems relying on two banks of nine batteries to lift the 85-foot, two-section Basque wooden ladder. They were considered to be very innovative, and both the hoist and turntable were powered by the batteries. These were also the first San Francisco ladder trucks with Hill-Table leveling devices. One of the units is now at the San Francisco Fire Department's museum. PAUL DARRELL.

Long Beach, California, bought this 1922 Ahrens-Fox Model K-S-4, serial number 1128, after it had been shown at the International Fire Engineers' Convention in San Francisco. It has been restored and can be seen at musters. The spherical dome in front is a single air chamber for the pump. WAYNE SORENSEN.

Aberdeen, Washington, purchased this 1927 Ahrens-Fox Model N-S-4, serial number 1676, powered by an Ahrens-Fox six-cylinder engine with a 5½-inch bore and a 7-inch stroke. For a time, it had been the Ahrens-Fox demonstrator in the Pacific Northwest. Because of the pump in front, Ahrens-Fox engines often ran hot. The firm designed its hood so the side flaps could fold under, opening the sides of the engine for fresh air. The rig was in active service until 1961 and in reserve status until 1969; it has since been restored. Note the white sidewalls, unusual on fire apparatus. WAYNE SORENSEN.

Motorized Fire Apparatus of the West • 93

Portland, Oregon, sold this 1928 Ahrens-Fox tractor (with a horse-drawn American-LaFrance 75-foot aerial ladder) to Beaverton, Oregon. The tractor's serial number was 2033. BOB ALLEN.

This is Tucson, Arizona's 1928 Ahrens-Fox 750-gpm pumper, serial number 1750. GEORGE H. SORENSEN.

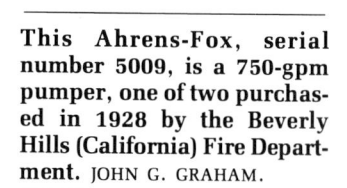

This Ahrens-Fox, serial number 5009, is a 750-gpm pumper, one of two purchased in 1928 by the Beverly Hills (California) Fire Department. JOHN G. GRAHAM.

94 • *Motorized Fire Apparatus of the West*

Above: This is a smaller — and less costly — Ahrens-Fox, produced in 1930 on a Schacht chassis with a Hercules motor. It was known as the Model V. This 500-gpm pumper, serial number 7009, was purchased by Menlo Park, California. WAYNE SORENSEN.

Below: Eureka, California, bought this 1936 Ahrens-Fox pumper (serial number 9008). Note the light color. It is now being restored. PAUL DARRELL.

Motorized Fire Apparatus of the West • 95

This 1938 Ahrens-Fox, serial number 9050, went to Long Beach, California. It had a 500-gpm pump, deck gun and full cab. DALE MAGEE.

Thermopolis, Wyoming, bought this 1938 Ahrens-Fox 500-gpm pumper, serial number 9015. FIRE ENGINEERING.

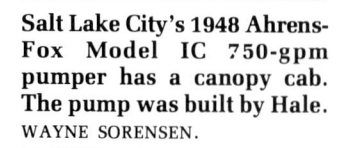

Salt Lake City's 1948 Ahrens-Fox Model IC 750-gpm pumper has a canopy cab. The pump was built by Hale. WAYNE SORENSEN.

96 • *Motorized Fire Apparatus of the West*

Portland, Oregon, used this 1911 American-LaFrance chemical and hose wagon; it was powered by a four-cylinder, 70-hp motor. Note the acetylene gas reflector headlights. DALE MAGEE.

American-LaFrance
Elmira, New York

This is probably the best-known of all United States fire apparatus manufacturers, and dates from before the turn of this century when a number of well-known manufacturers were joined in an attempt to dominate the industry. Throughout its long life, American-LaFrance built both distinctive custom apparatus as well as outfitted commercial chassis.

In 1903 Asa LaFrance designed, patented and built a spring-raised aerial truck. In the same year, the firm produced a number of steam-propelled chemical and hose wagons as well as numerous passenger autos. In 1904, after another reorganization, the American-LaFrance firm name was adopted. In 1905 the firm built its first motor-driven fire apparatus.

During 1909, the firm purchased an especially-designed four-cylinder Simplex engine, and then, in 1912, American-LaFrance designed its own six-cylinder engine with many features that made it more dependable than gasoline engines of that era. On American-LaFrance custom units, the engines had dual ignitions, two spark plugs per cylinder and dual

Oroville, California's first motorized apparatus was this 1912 American-LaFrance Type 10 chemical and hose car. The rig has been restored. WAYNE SORENSEN COLLECTION.

Motorized Fire Apparatus of the West • 97

This was originally a 1913 American-LaFrance Type 12 rotary gear 500-gpm pumper. Later in its life, the Logan (Utah) Fire Department rebuilt it into a 55-foot aerial ladder truck by extending its frame and adding a Pirsch Junior aerial ladder. WAYNE SORENSEN.

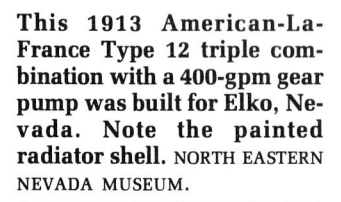

This 1913 American-LaFrance Type 12 triple combination with a 400-gpm gear pump was built for Elko, Nevada. Note the painted radiator shell. NORTH EASTERN NEVADA MUSEUM.

Berkeley, California's fire department used this Type 22, 1914 American-LaFrance front drive, chain-driven, two-wheel tractor, to pull a horse-drawn steamer. It is shown pumping at a general alarm fire on Shattack Avenue on September 17, 1923. BERKELEY FIRE DEPARTMENT.

1914 American-LaFrance Type 26 85-foot aerial ladder (straight frame with front-wheel-drive) was used by the Panama-Pacific Exposition's own fire department. (The exposition was held in San Francisco during 1914 and 1915.) After the exposition closed, the rig was purchased by the Fresno (California) Fire Department. Later, the unit was sold to Salinas, California, and the steering wheel moved to the left side. More recently, it was donated to San Jose (California) Fire Department and is being restored. Note the life net. WAYNE SORENSEN.

fuel systems with both a mechanical and electrical pump. (This is "redundancy" as is practiced today in the design of aircraft.) Prior to World War I, the firm also outfitted commercial chassis, especially Fords and Brockways.

In 1933, American-LaFrance and GMC worked together to build a new series of smaller pumpers for use in small communities. These rigs were powered by Buick engines.

On its custom rigs, American-LaFrance introduced 12-cylinder engines, enclosed cabs and metal ladders. After World War II, the firm introduced its 700 series, with the driver sitting ahead of the engine. American-LaFrance 700 series apparatus became the backbone of fire departments throughout the country.

American-LaFrance, a Figgie International company, is now building apparatus in Bluefield, Virginia.

Fresno, California, operated this 1915 American-LaFrance Type 19, with a 1,000-gpm rotary pump. The pump is behind the driver's seat; note the large air chamber. Observe also the twin or tandem hoods, necessary for the large, 150-hp, six-cylinder motor. FRESNO FIRE DEPARTMENT MUSEUM.

This is a 1915 American-LaFrance "Junior" Type 10, 300-gpm pump, used by Glendora, California. Junior pumps were less costly and designed for small town departments that wanted more dependability than most regular commercial chassis offered. This rig also had a 40-gallon chemical tank. It has been restored. BOB ALLEN.

Pocatello, Idaho's first motorized rig was this 1915 American-LaFrance Type 12 pump and hose car. About 20 years later the rig was "modernized" by having a windshield and electric flood light added. It also received a new engine (a Buda with 106 hp), new headlights and an electric siren. Lastly, a square gasoline tank replaced the original oval one. WAYNE SORENSEN.

American-LaFrance made only a few piston pumpers, and the only one delivered in the West went to Oakland, California. It was a 1917 Type 37, rated at 650 gpm. The unit today is at South Lake Tahoe, California. PAUL DARRELL.

100 • *Motorized Fire Apparatus of the West*

This is a 1917 American-LaFrance Type 17-6 tractor attached to a 1908 Seagrave city service trailer. It ran as Truck 17 for the San Francisco Fire Department. JOHN G. GRAHAM.

This started out as a 1916 American-LaFrance Type 17-6 tractor, used by the Portland (Oregon) Fire Department. After an accident, it was rebuilt. Observe the solid rubber tires on the rear of the tractor. The truck still exists. BOB ALLEN.

Seattle, Washington's 1919 American-LaFrance 1,000-gpm pumper after the fire department's shops rebuilt it in 1939. Some changes were left-hand steering, a Hall-Scott engine and Kenworth front end. DICK SCHNEIDER.

Motorized Fire Apparatus of the West • 101

Rawlins, Wyoming, had this 1923 American-LaFrance Type 31-4 straight-framed, rear-tillered aerial ladder truck. The ladder was raised by compressed coil springs, and its raising speed was controlled by an air cylinder and hand-operated brake. DAN MARTIN.

The Los Banos (California) Fire Department still has this 1924 American-LaFrance Type 39 triple combination pumper (serial number 4536) with a 600-gpm gear pump. The suction hose is connected and carried in a "squirrel-tail" position. It's powered by a 70-hp, six-cylinder motor. Note that the suction inlet is above the frame. WAYNE SORENSEN.

Kentfield, California, used this 1928 American-LaFrance Type 91 500-gpm pumper. Note the two-piece cast radiator. The pumper is still used as a parade piece. WAYNE SORENSEN.

102 • *Motorized Fire Apparatus of the West*

Nampa, Idaho's 1930 American-LaFrance Series 200 quad with a 1,000-gpm rotary gear pump. It is now in a Nampa museum. WAYNE SORENSEN.

1937 American-LaFrance Series 400, with a 1,000-gpm centrifugal Buffalo pump, was used by San Jose Fire Department. It was a Model 412-CB, serial number 7758, and powered by a V-12 engine. Notice the location of pump, just ahead of the cowl. The rig is still owned by the San Jose Fire Department. DON WOOD.

This smaller American-LaFrance pumper was designated the Scout, and was one of two purchased in 1937 by Petaluma, California. It was powered by a Lycoming eight-cylinder engine, and had a 500-gpm gear pump and small water tank. WAYNE SORENSEN.

Motorized Fire Apparatus of the West • 103

1938 American-LaFrance, serial number 7823, 65-foot water tower and combination service truck was delivered to Los Angeles. Powered by a V-12 250-hp engine, the truck was both the only one of its kind built and the last tower constructed in the United States. AMERICAN-LaFRANCE.

San Rafael, California, bought this 1938 American-LaFrance quad with a 750-gpm rotor gear pump. It was powered by a 250-hp V-12, and carried a 75-gallon booster tank. WAYNE SORENSEN.

One of four American-LaFrance Duplex pumpers delivered to Los Angeles Fire Department in 1938. Los Angeles Fire Chief Ralph J. Scott designed them to reduce the number of engines working a major fire. Each Duplex could take the place of three conventional pumpers. Each rig carried two engines and two pumps, and they responded with special manifold wagons (shown later in this chapter with Seagrave apparatus). This was Engine Company 17's pump. DALE MAGEE.

104 • *Motorized Fire Apparatus of the West*

Reno, Nevada, purchased this 1941 MLF Series 500 100-foot aerial ladder. The cab could carry seven. The V-12 engine produced 240 hp. WAYNE SORENSEN.

In 1941, the United States Navy installation at Mare Island, California, received this American-LaFrance 100-foot aerial. It was powered by a 190-hp V-12 engine. The straight frame-forward control design allowed for a much shorter overall truck. PAUL DARRELL.

This is San Jose (California) Fire Department's Engine No. 4, a 1944 American-LaFrance 750-gpm pumper working at the Cheim Lumber Company and Pacific Hardware and Steel Company fire on April 15, 1955. Heat from the fire blistered the truck's paint. SAN JOSE FIRE DEPARTMENT.

Motorized Fire Apparatus of the West • 105

This 1949 American-LaFrance 700 Series 100-foot aerial ladder truck, was owned by the Denver (Colorado) Fire Department. It had a five-man half cab. The 700 Series was very popular, and eventually all other manufacturers offered a cab-ahead-of-engine design. DICK ADELMAN.

In 1960, American-LaFrance built this turbine-powered 900 Series 1,000-gpm pumper for the San Francisco Fire Department. It was powered by an experimental Boeing Turbo-jet gas turbine 352-hp engine. (Seattle received a 100-foot aerial with similar power at the same time.) San Francisco's rig ran as Engine No. 14. After three years, it was decided that the power plant was unsatisfactory; it was unreliable, had sluggish acceleration and took over a minute to warm up. Also, heat from the overhead exhaust (see picture) could ignite overhead awnings or wires. The power plant was replaced with a conventional Continental six-cylinder 330-hp engine. AMERICAN-LAFRANCE.

This experimental gas turbine engine was used in San Francisco's 1960 American-LaFrance. BOEING AIRPLANE COMPANY.

106 • *Motorized Fire Apparatus of the West*

American-LaFrance also manufactured commercial trucks intermittently from 1913 until 1929. This 1926 Model Y American-LaFrance was used by the Los Angeles factory branch to service American-LaFrance products in Southern California. BILL WEST.

American-LaFrance outfitted this 1915 Model T Ford with chemical tanks for Burbank, California. American-LaFrance outfitted many Fords, and rigs like this were designated as "American-LaFrance Type 32." WAYNE SORENSEN.

American-LaFrance offered a series of light-duty apparatus on Brockway chassis. (Brockways were built in Cortland, New York, near American-LaFrance's Elmira, New York, headquarters.) The nameplates read: "LaFrance-Brockway Torpedo." This 1923 model, Type 36, with a 250-gpm pump was purchased by the Atwater (California) Fire Department. The unit is still used in parades. WAYNE SORENSEN.

Motorized Fire Apparatus of the West • 107

This is a 1927 American-LaFrance Cosmopolitan triple combination Type 46 with a 400-gpm pump on a Brockway chassis. It's powered by a Wisconsin motor, and belonged to Kingsburg, California. PAUL DARRELL.

California's San Quentin Prison received this 1928 GMC-American-LaFrance, which is shown undergoing its acceptance tests. It had a 500-gpm gear pump. WAYNE SORENSEN.

When the long Bay Bridge between Oakland and San Francisco opened in 1936, this 1936 GMC-American-LaFrance rig was used for fire and rescue work. It had a 500-gpm pump, and later was sold to Legget, California. PAUL DARRELL.

108 • *Motorized Fire Apparatus of the West*

Above: Rear view of a 1931 GMC-American-LaFrance 500-gpm rotary gear pumper used by the Hemet (California) Fire Department. The rig exists today. DON WOOD.
Below: Close-up of grille emblem shows both GMC and American-LaFrance. DON WOOD.

Motorized Fire Apparatus of the West • 109

This is a streamlined-style Buffalo with ladders, and even the pump panel, enclosed. This 1940 500-gpm pumper belongs to the Winters (California) Fire Department. WAYNE SORENSEN.

Vallejo, California, used this ex-United States Navy 1942 Buffalo, which had a 750-gpm pump (behind the door), a 200-gallon water tank and an overhead ladder rack. WAYNE SORENSEN.

Wartime production and material restrictions show on this Buffalo 750-gpm pumper, one of two that went to the El Cerrito (California) Fire Department. There is no chrome and no streamlined body panels (to save metal and labor). WAYNE SORENSEN.

110 • Motorized Fire Apparatus of the West

This 1946 Buffalo 750-gpm pumper went to the Jerome (Idaho) Fire Department. WAYNE SORENSEN.

Buffalo Fire Appliance Corporation
Buffalo, New York

The Buffalo (New York) Fire Extinguisher Manufacturing Company started out in 1920 by supplying fire engine bodies for mounting on commercial chassis. In 1927, they started building their own custom chassis as well, and soon became a major apparatus builder. They used Hale pumps ranging from 350 to 1,000 gpm and Hercules motors.

In 1937, they introduced a "limousine" style of apparatus, which enclosed virtually all of the equipment the rig carried. Even the pump panel was behind a cabinet door. Buffalo also built city service and ladder trucks with the ladder racks completely enclosed. The firm ceased production in 1949. During its life, a small number of its products reached the West Coast.

Ford dealers sold Buffalo equipment on Ford chassis. This 1935 Ford-Buffalo went to Battle Mountain, Nevada. WAYNE SORENSEN.

Motorized Fire Apparatus of the West • 111

Above: Engine Company 29 of the Los Angeles Fire Department attached this Christie tractor to pull their steam pumper. CLANCY CRUM COLLECTION.

Below: This Christie tractor was used by the Tacoma (Washington) Fire Department to pull a 1907 Nott 800-gpm steam pumper. It ran as Engine No. 1. RALPH DECKER COLLECTION.

A scene which took place at every firehouse was the arrival of a motorized rig to take the place of horses. Here, on September 19, 1915, in front of the San Francisco Fire Department's Engine Company No. 20 house, a 1915 Christie is attached to a 1908 Metro 700-gpm steam pumper, serial number 3250. On the left is a 1914 White chemical and hose wagon. SAN FRANCISCO FIRE DEPARTMENT.

Christie Front Drive
Hoboken, New Jersey

This firm was one of several which addressed the problem of supplying a tractive power to replace horses in front of steamers and ladders. In 1911, John Walter Christie developed a two-wheel auto-tractor for the New York City Fire Department. This tractor had a four-cylinder engine mounted longitudinally, ahead of the axle. The unit had a two-speed gearbox from which power was transmitted by chain to a countershaft, at the end of which were spur pinions that meshed with gear inside the road wheels.

Christie tractors built after 1912 had transverse engines. Some tractors were started by hand cranking, others had compressed air starters. Christie ended production in 1918 after building about 600 tractors. Western cities that used Christie tractors were Denver, Colorado; Fresno, Los Angeles, Oakland and San Francisco, California; Portland, Oregon; Seattle and Tacoma, Washington.

W.S. Darley
Chicago, Illinois

This company was formed by W.S. Darley, an inventor, in 1908. The firm produced both pumps and

Beaverton, Oregon, added this rakish ragtop to their 1935 Chevrolet-Darley, which was equipped with a 500-gpm Champion midship centrifugal pump. BOB ALLEN.

Motorized Fire Apparatus of the West • 113

A familiar site around United States military bases during World War II were Darley-outfitted pumpers, known as Class 325. This one, on a Chevrolet chassis, served at the Veterans Administration Hospital in Boise, Idaho. WAYNE SORENSEN.

This 1952 International was outfitted by Darley for use by the United States Army with a 500-gpm pump and a 150-gallon booster tank. Later, the rig was used by the Benicia (California) Fire Department. WAYNE SORENSEN.

Darley placed a 500-gpm front-mount Champion pump on this 1960 Ford tilt-cab. The rig was used by the Cole Fire District near Boise, Idaho. WAYNE SORENSEN.

114 • Motorized Fire Apparatus of the West

FWD built this open-seat Model KHS 500-gpm pumper for Anchorage, Alaska, in 1942. The 300-gallon water tank, ladders and suction hose are all enclosed. FWD.

apparatus, and it competed with Barton front-mounted pumpers for the low-cost fire engine market. For example, in 1934 Darley advertised a complete Ford-Darley 500-gpm pump for under $2,000. The firm has always viewed small, volunteer fire departments with limited budgets as its best market.

The Darley pump was called the "Champion" and included both midship and front-mount models. Champion pumps were also sold to other apparatus outfitters. During World War II, Darley had a large contract to build front-mount high-pressure fog pumps for the United States Army Air Corps. At about the same time, Darley introduced the first three-stage centrifugal pump which could simultaneously provide high-pressure fog through booster hose, plus a high volume of water through 2½-inch hose lines.

The firm is now located in Melrose Park, Illinois.

FWD
Clintonville, Wisconsin

Otto Zachow is the inventor upon whom the success of this firm is based. He invented a double Y universal joint encased in a ball-and-socket that allowed power to be applied to the front driving wheels of a vehicle and still permit the wheels to be steered.

The firm built both autos and trucks, although it is noted primarily for its all-wheel-drive trucks. In

Built for the Salt Lake City Fire Department, this 1948 FWD Model KSU has a closed cab and a 750-gpm pump. FWD.

Motorized Fire Apparatus of the West • 115

Benicia, California's 1950 FWD, has a 750-gpm pump and a chromed radiator shell and grille. It is still in service. PAUL DARRELL.

In 1953, FWD delivered this 100-foot aerial to Beverly Hills, California. Note the wood-trussed aerial ladder. The beams are made of straight-grain Douglas fir, the rungs of straight-grain oak or ash. The weight of the ladder is carried by the trusses. At that time, most aerial ladders were metal as they are now. BOB ALLEN.

Sometimes, large firms purchase fire apparatus to protect their own property. The Union Pacific Railroad bought this 1953 FWD with a Waterous 500-gpm pump to use in their railroad yard at Pocatello, Idaho. WAYNE SORENSEN.

116 • Motorized Fire Apparatus of the West

This is a 1959 FWD high-pressure wagon used by Los Angeles. Interestingly, it is not all-wheel-drive. It's powered by a Hall-Scott engine and equipped with two Waterous booster pumps, which could deliver through all six outlets. The water battery was a "Morse Invincible" with 1¾-, 2- and 2¼-inch tips. The rig ran as part of a two-piece engine company. BOB ALLEN.

1914, the first FWD fire engine was built, using a Northern pump. World War I did much to help the firm, mainly because of large orders and the fine reputation that FWDs earned in war service.

After the war, the firm expanded both its commercial and fire apparatus lines. Many Wisconsin communities bought FWD equipment, and Chicago, Illinois, was one of its best customers. The New York City Fire Department was also a big user, and by 1936 they owned 68 pieces of FWD apparatus. The company's history proudly tells of a new order they received that same year from New York City for 12 aerial ladder trucks. "Specifications called for a special tractor and semi-trailer, nearly 62 feet long in overall length, powered by a 150-horsepower engine, carrying 339 feet of ladders, among them an 85-foot suspension ladder, capable of being elevated and raised in less than 20 seconds to a full vertical position with a 200-pound weight on the extreme tip."

FWD fire apparatus can be found in the West, especially in situations where its superior traction is needed. In 1963 FWD acquired Seagrave (which is mentioned later in this chapter). The firm is in business today.

Hale Fire Pump Company
Conshoncken, Pennsylvania

This firm began prior to World War I in Wayne, Pennsylvania, and the first pumper it built was on a Simplex chassis. Between the two world wars, the firm built a number of its units on commercial chassis. In 1940, they decided to build fire pumps only, and have done so to the present.

Hale Fire Pump Company sold this 1924 pumper to the Lake Shore Fire Department, near Clear Lake, California. WAYNE SORENSEN.

Motorized Fire Apparatus of the West • 117

A 1917 Ford Model T equipped with a Buckeye Manufacturing Company piston pump. The pumper was distributed by Howe Fire Apparatus Company and this one purchased by Turlock, California. The completely restored pumper is now a parade piece. PAUL DARRELL.

This 1928 Ford-Howe combination chemical and hose car was built for Leadville, Colorado. WESLEY HAMMOND.

1929 Howe Defender on a Chevrolet chassis had a 500-gpm rotary gear pump and a 500-gallon water tank. It was sold to Almeda, Idaho, now a part of Pocatello. WAYNE SORENSEN.

118 • *Motorized Fire Apparatus of the West*

Pictured here is a 1931 Howe Custom Defender on a Gramm chassis, with a Waterous centrifugal positive displacement 750-gpm pump, at Boise, Idaho. Later, it was sold to nearby Eagle. HANK GRIFFITHS.

Howe
Anderson, Indiana

In 1872, B.J.C. Howe introduced the first piston-type pump for fire department use. The pump could be operated by a team of 20 men or a team of horses. In its early days, the firm was located in Indianapolis, Indiana. It started building horse-drawn gasoline motor pumps (and many of these units came with spare handles for manual use, in case the gasoline motor failed).

In 1906, Howe built his first automobile pumper, which was sold to LaRue, Ohio. It had a 250-gpm pump, which had levers for manual use in case of motor failure. In World War I, Howe had a government order for 163 pumps to protect military bases.

Howe moved from Indianapolis to Anderson in 1917. This was to be closer to the Buckeye Company, which was developing the Lambert automobile. Howe wanted to use the Lambert chassis for mounting his fire apparatus. This association lasted for only a few years; apparently little ever came of the Lambert.

Howe continued to place his pumps and bodies on many makes of commercial chassis, including the Ford Model T and TT. In the 1920s, Howe introduced his custom Howe Defender, which was built on a Defiance chassis. When Defiance dropped out of business in 1930, Howe used his own chassis for the Defender. More recently, the Defender was built on Duplex chassis,

In 1974, Howe acquired the Oren Roanoke Corporation and Coast Apparatus, Inc. (see Chapter Two). More recently, Howe has become part of Grumman Emergency Products.

In 1931, Boise, Idaho, purchased this Studebaker-Howe booster and squad truck, which was equipped with a 300-gpm centrifugal pump and water tank. HOWE.

Howe built this 750-gpm centrifugal pumper on a Sterling chassis for the United States Army in 1942. Later, it was acquired by Dillon, Colorado. ROLAND BOULET COLLECTION.

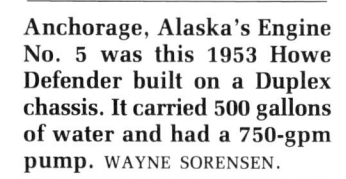

Anchorage, Alaska's Engine No. 5 was this 1953 Howe Defender built on a Duplex chassis. It carried 500 gallons of water and had a 750-gpm pump. WAYNE SORENSEN.

Pollock Pines, California, runs this 1959 International-Howe. It carries 1,000 gallons of water and has a front-mount 750-gpm pump. WAYNE SORENSEN.

120 • Motorized Fire Apparatus of the West

Above: Shortly after World War II, Howe began mounting fire fighting equipment on Jeeps. Idaho City, Idaho, purchased this 1960 model, which has a 400-gpm Barton pump and a 70-gallon water tank. WAYNE SORENSEN.

Below: Rincon Valley, California, operates this 1960 Willys FC 170 Howe with a 400-gallon water tank and a 500-gpm centrifugal pump. BOB ALLEN.

WHAT THEY THINK OF
Knox APPARATUS
ON THE PACIFIC COAST

Vallejo Evening News

VALLEJO, CAL.   FRIDAY, FEB. 6, 1914

Knox Wagon Finest Piece of Fire Apparatus on Pacific Coast

In the new Knox motor-driven hose and chemical wagon built for the City of Vallejo by the Reliance Automobile Company, the municipality has one of the finest pieces of fire apparatus to be found on the Pacific coast. Yesterday afternoon a representative of the News was present when the machine was given one of its official tests and it certainly exceeded all expectations.

Loaded down with 1,200 feet of hose, both chemical tanks filled and four men riding on it, the apparatus went up the steepest hills in town without any trouble whatever. Starting from the corner of York and Marin streets the wagon went to the top of York street hill in 40 seconds. and then came back and went from the corner of Sacramento and Virginia streets to the top of Virginia street hill in 45 seconds. On another trip up Virginia street hill the engine was killed on the steepest part of the grade and the apparatus allowed to come to a dead stop, after which it was started up again and went up the grade without any trouble.

After performing every conceivable stunt on the grades in town, Demonstartor Gibbons took the apparatus out on the Napa road and a run was made, a speed of 48 and 50 miles an hour being made on the county road. The big car runs as smooth as a sewing machine and it is a pleasure to ride on it.

From the performance of the apparatus it would seem that Vallejo should be well protected from the ravages of fire in future. The motorwagon would be at the scene of a fire before the horse-drawn appartus could get out of the fire department headquarters. This modern firefighting apparatus and the splendid supply of water in the municipal lakes should be a great factor in securing a reduction in insurance rates in Vallejo.

KNOX DOUBLE TANK COMBINATION CAR

Send for circulars and list of cities using Knox Motor Driven Apparatus

KNOX AUTOMOBILE CO.

FACTORY AND MAIN OFFICE: SPRINGFIELD, MASS.

NEW YORK   BOSTON   CHICAGO   SAN FRANCISCO   ST. LOUIS   PITTSBURG   KANSAS CITY

This 1914 advertisement was a testimonial to Vallejo, California's Knox. FIREMEN'S HERALD.

122 • *Motorized Fire Apparatus of the West*

The early Knox automobile gained the fire service's confidence, and was a pioneer in motorized fire apparatus. This 1914 Knox combination hose wagon and chemical engine was operated in Berkeley, California. BERKELEY FIRE DEPARTMENT.

Knox and Knox-Martin
Springfield, Massachusetts

The Knox Automobile Company was founded in 1898, and was one of the first firms to manufacture a water-cooled automobile. Knox's first fire apparatus was delivered to the Springfield (Massachusetts) Fire Department in 1906. Knox developed a full line of apparatus, and the firm promoted its products by mailing picture postcards containing specifications and prices to fire chiefs throughout the United States.

In 1913 Knox built a piston pumper for Springfield. During 1914, Willows, California, purchased a Knox piston pumping engine. The apparatus featured a body known as the "Springfield-style" body, which had a bench seat for the firemen. Several other Knox units were sold in the West.

In 1909, a former Knox employee, Charles Hay Martin, returned to the firm to become its chief engineer. Martin developed a system for attaching tractors to trailers, which was known as the "Martin Rocking Fifth Wheel." The turntable was carried by semi-elliptical springs attached directly to the rear axle of the tractor, so that the weight of the trailer was carried by the tractor's rear wheels, while the much lower weight of the tractor itself was carried by separate, lighter capacity springs. (This system is still used today.)

The first Knox-Martin tractors were three-wheelers, powered by 40-hp, four-cylinder engines. These tractors were used to pull steamers, aerials and water towers. The Knox-Martin tractor's steering column extended over the hood to a steering box above the single front wheel. The front wheel could turn near-

Denver, Colorado's 1914 Knox-Martin tractor, pulling a 1909 horse-drawn Seagrave 85-foot aerial (serial number 3633). This unit was employed as late as World War II, when it was used at Lowry Field, Colorado, for changing light bulbs inside aircraft hangers. ROBERT SAMS.

Motorized Fire Apparatus of the West • 123

The fire department at Willows, California, purchased this 1914 Knox piston pump. It had a six-cylinder water-cooled engine, which drove a 600-gpm three-piston pump. The body design is referred to as the "Springfield style" hose body, meaning it was flared and had a full seat above it for firemen. WILLOWS FIRE DEPARTMENT.

San Jose, California, had three Knox-Martin tractors. This one still exists and is shown with an American-LaFrance steam pump (serial number 2676) that was acquired by San Jose from the San Francisco Fire Department. The steam pump saw service during San Francisco's 1906 earthquake and fire as Engine 28. The tractor-steamer combination is now used in parades and musters. WAYNE SORENSEN.

View of a 1920's Mack semitrailer Type AC-7 showing the rear steering. It is worm-and-gear-type; the rear axle of the trailer is equipped with pivot-type steering knuckles. The tiller shaft and tiller wheel are mounted in the center. MACK MUSEUM.

124 • *Motorized Fire Apparatus of the West*

Boise, Idaho, used this 1923 Mack AC four-cylinder tractor to replace a 1911 Seagrave tractor for pulling their 65-foot aerial. This combination saw service until 1941. HANK GRIFFITHS.

ly 90 degrees in either direction, allowing the tractor to turn a little more than its own length. A number of these Knox-Martin tractors were sold in the West. Their normal road speed was about 10 mph, but by altering sprockets as would be done for the fire apparatus, this could be increased to over 30 mph.

Mack
Allentown, Pennsylvania

The five Mack Brothers founded this company in Brooklyn, New York, at the turn of this century. They built commercial buses and trucks. In 1905 they moved to Allentown, Pennsylvania. During 1909 Mack built a tractor for the Allentown Fire Department to use to pull an aerial. In 1911 the firm built its first pumper. Mack was — and is — primarily a manufacturer of commercial trucks, and fire apparatus has been but one aspect of their business. Some early Mack developments, important to fire apparatus, dealt with heating and cooling of engines, important for pumpers.

In 1914, Mack introduced the AB model truck chassis, which was widely used for carrying pumpers, combination chemical and hose cars and city service trucks. The more famous AC, or "Bulldog" (World War I soldiers gave it this name because of its pugnacious appearance), model was introduced in 1916 and continued in production until nearly the beginning of World War II. The AC Bulldog saw wide use in the fire service.

During this period, Mack decided to market their own apparatus. Using a pump from Northern Fire Apparatus of Minneapolis, Mack engineers developed their own "Mack" pumper. In 1927 Mack brought out the first series-parallel, pressure-volume centrifugal pump that was to set the standards for pumping efficiency. In that same year, they also adopted four-wheel brakes on their apparatus.

Left view of an early 1930's Mack Type 75 pumper under construction. It has a midship pump and a Mack CT six-cylinder engine developing 140 hp. MACK MUSEUM.

Motorized Fire Apparatus of the West • 125

One of two identical fireboat tenders built on 1925 Mack bus chassis for Los Angeles. These tenders were assigned to harbor areas and ran with fireboats. They assisted in situations where the fireboats' massive pumps and towers could not effectively reach the fire. These tenders carried 3½-inch hose and mounted a heavy-turret nozzle. BOB ALLEN.

In 1926 Mack introduced a new series of fire engine having the AP six-cylinder, 150-hp engine. The first one was built for Seattle, Washington. The Byron-Jackson pump, made in Berkeley, California, was a 1,000-gpm series-parallel multi-stage centrifugal. The truck had chain drive, four-wheel brakes and electric starting. MACK.

San Francisco had two of these 1928 Mack Bulldog Model AC water towers, powered by four-cylinder, 70-hp motors. The 35-foot Gorter towers were built and installed by the Union Iron Works of San Francisco. This rig is now in the San Francisco Fire Department Museum. WAYNE SORENSEN.

126 • Motorized Fire Apparatus of the West

San Francisco's Fire Department purchased two Mack type 15 model AP rigs, one in 1928 and one in 1929. They had Byron-Jackson parallel series 1000-gpm centrifugal pumps and had worm-gear drive. WAYNE SORENSEN.

In 1928, Mack introduced an engine-driven mechanical hoist for aerial ladders. The truck's power take-off was used. Power was transmitted via a vertical shaft that passed directly through the center of both the tractor's fifth wheel and the aerial ladder's turntable. This was an important development because up to that time ladders were raised by springs, compressed air, and human muscle.

As other Mack truck models were introduced with newer styling, the appearance of the Mack fire apparatus would be simultaneously updated. In the 1930s, Mack pioneered in the adoption of the sedan cab, which provided better protection for the fire fighters.

Mack continued to sell its chassis to other apparatus outfitters as well as build a full line of custom apparatus. In 1984 the firm dropped out of the custom apparatus field because of high per-unit engineering and production costs. They decided to concentrate their efforts on supplying chassis to other apparatus suppliers. In 1987, they moved most of their operations to South Carolina.

The Mack AL chassis was designed for buses, but here's an example of one being used for fire apparatus. This 1928 Mack Type 90, with a 1,000-gpm Byron-Jackson pump and 90-gallon water tank, was used by the Ross (California) Fire Department. The vehicle is now in private hands. WAYNE SORENSEN.

Motorized Fire Apparatus of the West • 127

Burbank, California, used this 1928 Mack AL Type 90 combination pumper with a 1,000-gpm Byron-Jackson pump and two chemical tanks. The running board has a hinged section to allow access to the suction port. Note the truck's low lines. The AL styling managed to smooth some of the Bulldog's rugged features. BURBANK FIRE DEPARTMENT.

This is a Bulldog quadruple service truck with the Model L six-cylinder engine, and was delivered to Coronado, California. It had a 500-gpm pump, the longest ladder was 50 feet. MACK.

Seattle, Washington's Type 90 Mack tractor-drawn 85-foot aerial was built in 1929. It is equipped with an aerial ladder hoist driven by the truck's engine. The rig was repowered with a Hall-Scott 400 engine in 1937. It is now part of the Seattle Last Resort Fire Department. BOB ALLEN.

128 • Motorized Fire Apparatus of the West

This 1930 Mack 1,000-gpm pumper was used by Carmel, California. The hood has been lengthened to enclose a Hall-Scott motor. WAYNE SORENSEN COLLECTION.

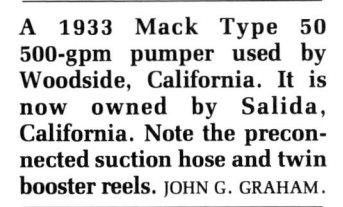

A 1933 Mack Type 50 500-gpm pumper used by Woodside, California. It is now owned by Salida, California. Note the preconnected suction hose and twin booster reels. JOHN G. GRAHAM.

Stockton, California, used this 1933 Mack Type 90 1,000-gpm quad. The truck was loaded down with a total of 17 ladders, totaling 382 feet in length. RICHARD A. COWAN.

Motorized Fire Apparatus of the West • 129

This powerful-looking 1934 Mack with a 1,000-gpm pump was delivered to Kellogg, Idaho. Note the mechanical louvers on the radiator. KELLOGG FIRE DEPARTMENT.

The San Marino (California) Fire Department used this 1935 Mack Type 75, 750-gpm quad. Observe the Bulldog emblem on the radiator cap. The quad was powered by a 140-hp Mack CT six-cylinder engine. DALE MAGEE.

Phoenix, Arizona, used this 1935 Mack with a 1,000-gpm pump. WAYNE SORENSEN COLLECTION.

Stockton, California, purchased this 1938 Mack Type 80 city service truck. Note the new styling, similar to Mack's commercial trucks. The rig's Mack Thermodyne engine was rated at 168 hp. STOCKTON FIRE DEPARTMENT.

The Owyhee (Nevada) Volunteer Fire Department used this Navy surplus 1941 Mack Type 75, 750-gpm pumper. A 180-gallon tank was installed ahead of the hose box. WAYNE SORENSEN.

This 1943 wartime Mack Type 95 1,000-gpm pumper was delivered to Pocatello, Idaho. It had a 160-gallon water tank. Pocatello added the overhead ladder rack in 1947. WAYNE SORENSEN.

Payette, Idaho's 1948 Mack 750-gpm, Type 75 pumper with a 500-gallon water tank and enclosed overhead ladder rack. WAYNE SORENSEN.

This 1958 Mack B-22 1,500-gpm pumper was used by Seattle. The pump was a Hale two-stage midship QF. The rig was powered by a Hall-Scott 323. It also had a 200-gallon booster tank. Notice the recessed headlights. It ran as Engine No. 38. DICK SCHNEIDER.

Los Angeles used this 1958 Mack C pumper. It had a Hall-Scott motor, a 1,250 gpm Hale pump, and a Morse turret. WAYNE SORENSEN COLLECTION.

The L.N. Curtis & Sons firm (discussed in Chapter Two) was Maxim's West Coast distributor after World War II, and here's a 1948 rig Curtis sold to Glenns Ferry, Idaho. It had a 750-gpm pump and a 500-gallon water pump. Notice the turnout coats hanging on the side, ready for volunteers when they reached the blaze. WAYNE SORENSEN.

Maxim Motor Company
Middleboro, Massachusetts

The first Maxim motorized fire apparatus was built in 1914. It was a chemical and hose car on a Thomas chassis. The firm had been founded by Charlton Maxim, the one-time Middleboro (Massachusetts) fire chief who had gained some experience in building a motorized rig for his own department. More orders arrived, and a new business was on its way.

Maxim expanded its line of fire apparatus and also built trucks for specialized commercial uses. The firm's fire apparatus maintained a fine reputation, and some units were marketed in the West. Maxim supplied aerial ladders to Crown and Mack.

The firm continues in business.

Orland, California, used this 1948 Maxim, which has a 500-gpm Hale pump. Note hard-suction hose wrapped around the front and roof-mounted red light and siren. ROLAND BOULET COLLECTION.

Motorized Fire Apparatus of the West • 133

Portland, Oregon's Engine No. 29 used this 1952 Maxim 1,500-gallon triple combination pumper. Note the double bumper bars, common to Maxims at that time. BOB ALLEN.

Redwood City, California, bought this 1953 Maxim straight-frame 75-foot aerial ladder truck. Maxim built a high-quality aerial ladder at this time and supplied them to several other apparatus builders. WAYNE SORENSEN.

Maxim outfitted a number of commercial chassis for use by the military during World War II. After the war, this 1942 Ford-Maxim 500-gpm pumper was acquired by Auburn, California. WAYNE SORENSEN.

134 • *Motorized Fire Apparatus of the West*

Before building custom apparatus, Pirsch outfitted chassis supplied by others. This is a 1913 Atterbury-Pirsch city service truck, which was used by the Aberdeen (Washington) Fire Department. It had chain drive, hard rubber tires and a chemical tank. Note the two "King" Dietz oil lanterns on the side. PIRSCH.

Peter Pirsch & Company
Kenosha, Wisconsin

Peter Pirsch, son of a pioneer Wisconsin wagon builder, in 1899 patented a compound trussed extension ladder. Pirsch had built early hand-drawn ladder trucks and then horse-drawn trucks. He also was a supplier of ladders and chemical equipment to other apparatus manufacturers. The first motorized ladder truck Pirsch built was on a Rambler chassis.

In 1916 Pirsch delivered his pumper, built on a White chassis, to Creston, Iowa. Soon he was building rigs on numerous commercial chassis. During 1927 Pirsch introduced a line of custom apparatus; a year later he built and sold the first closed-cab fire engine in the United States.

Four years later, in 1931, Pirsch made a major contribution to fire fighting. This was his patented hydromechanical aerial ladder hoist. This type of aerial could be easily controlled by one person. In 1932, the city of Spokane, Washington, purchased one of these units pulled by a White tractor. In early 1938 Pirsch delivered the first aluminum alloy aerial ladder of closed lattice design, the design is still being used. That same year Berkeley, California, purchased the first of these ladders to be used in the West.

The Pirsch firm continues in business. However, in October 1986 the firm announced that it was closing its "custom" building operation at Kenosha following a labor dispute, although, some apparatus is still built in Kenosha.

During World War II, Maxim built trailer-mounted pumps. They were used in defense plants throughout the U.S. and also shipped abroad. The insets show them being loaded aboard railcars. AMERICAN TRUCK HISTORICAL SOCIETY.

Motorized Fire Apparatus of the West • 135

Pirsch outfitted this 1926 Federal Big Six for Yerington, Nevada. The engine was a six-cylinder Continental, and the pump was rated at 500-gpm. WAYNE SORENSEN.

This is an early piece of Pirsch "custom" apparatus, a 1927 Model 17 750-gpm pumper delivered to Bend, Oregon. It was powered by a Waukesha engine. The rig has been restored. WAYNE SORENSEN.

An impressive Pirsch ladder truck, with a hydraulically-operated 65-foot metal aerial ladder, being unloaded from a railcar for service in Lewiston, Idaho. The serial number was 1057. HAROLD SCHAPELHOUMAN.

Above: This photo shows Berkeley, California's 1938 Pirsch (serial number 1044) after it was rebuilt — about 1968 — and its 85-foot aerial replaced with a 100-foot one. Note the windshield for the tillerman. The original 1938 rig was the first Pirsch aluminum aerial ladder delivered to California. WAYNE SORENSEN.

Below: This 1940 Pirsch quintuple unit, serial number 1167, was built for Anaheim, California. Its equipment included a 750-gpm Hale centrifugal pump, a 65-foot aluminum ladder, booster equipment, hose and ground ladders. BOB ALLEN.

Sacramento, California, bought this 1941 Pirsch 100-foot aerial ladder truck. It's now owned by the Sacramento Fire Buff Club. WAYNE SORENSEN.

In 1942, Pirsch constructed this Model 20G for the United States Navy. After World War II, it was acquired by the fire department in Yerington, Nevada. RICHARD A. COWAN.

Richmond, California, received this 1944 Pirsch Model 41, 65-foot three-piece intermediate aluminum hoist aerial. It was powered by a Waukesha 145 GK engine. WAYNE SORENSEN.

138 • Motorized Fire Apparatus of the West

Above: This 1946 Pirsch 1,000-gpm pumper with a two-man GMC cab went to Tacoma, Washington. In 1963, it was destroyed in a fire at Pier 7. RALPH DECKER.

Below: This is one of two 1947 Pirsch Model 38 city service trucks purchased by Oakland, California. City service trucks were rare in this era because aerial ladder units could perform all the functions of a city service truck, plus having an aerial ladder. Oakland liked Pirsch apparatus; between 1941 and 1953 they purchased 15 pieces. BOB ALLEN.

In 1950, Los Angeles bought two Pirsch Model 47E X series aluminum three-piece, 85 foot, aerial ladder trucks, powered by Hall-Scott 470 engines. GEORGE "SMOKY" BASS.

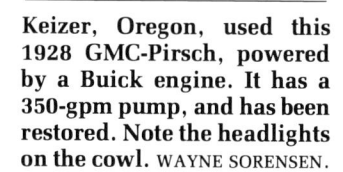

Keizer, Oregon, used this 1928 GMC-Pirsch, powered by a Buick engine. It has a 350-gpm pump, and has been restored. Note the headlights on the cowl. WAYNE SORENSEN.

Oakland, California's 1952 Mack-Pirsch with a 65-foot wooden aerial ladder. It was powered with a Hall-Scott engine. WAYNE SORENSEN.

140 • *Motorized Fire Apparatus of the West*

This 1956 White-Pirsch quintuple unit was built for Kentfield, California. The 85-foot Pirsch aerial ladder is made of compact aluminum alloy. The rig has a 1,000-gpm pump. Recently, it was repainted a lighter color. WAYNE SORENSEN.

Seagrave
Columbus, Ohio (now in Clintonville, Wisconsin)

This company was founded in Detroit in 1881 by Frederick S. Seagrave (being one of the first vehicle industries in what was to become "auto city"). Seagrave got into the business indirectly. Originally, he built ladders for use in Michigan's orchards. He was approached by neighboring volunteer fire departments for assistance in constructing carts for carrying ladders. Seagrave began building hand-drawn two-wheel carts, then four-wheel ones, and soon he started constructing horse-drawn ladder trucks. Before long, Seagrave had both feet in the fire apparatus business. In 1891, he changed the name of the firm to "The Seagrave Company" and moved his business from Detroit to Columbus where it was to gain nationwide recognition.

In 1907, Seagrave made its first motorized apparatus, a combination chemical and hose car powered by a four-cylinder air-cooled engine. In the same year, the firm delivered a hose wagon to Vancouver, British Columbia. These early Seagraves had what is referred to as "buckboard" styling because they looked so much like the horse-drawn buckboard. Seagraves of this era also had an ignition consisting of two independent sets feeding duplicate spark plugs for each cylinder; one set was powered by a storage battery, the other by high-tension magnets. Many of these Seagrave buckboards appeared in the West.

The company also went into the aerial ladder market. One of its first units was also delivered to Vancouver in 1907. Tacoma and Seattle (Washington) were also early users. Seagrave also introduced a "sidesaddle" form of aerial truck, with the driver's

This is Riverside, California's 1907 Seagrave chemical hose unit pulling an American-LaFrance 600 gpm steamer. WAYNE SORENSEN COLLECTION.

Motorized Fire Apparatus of the West • 141

Every bit as shiny as it was when new in 1909, is Pasadena, California's restored Seagrave AC 40 (air-cooled, 40-hp motor) chemical engine, serial number 3527. The rig came with two 60-gallon soda-acid tanks, a hose reel, two ladders, Rushmore carbide headlights and nine-inch flood light and a friction-drive siren. In 1923, it was converted to a flood light and high-pressure wagon and much of the original equipment was removed. It was kept on the Pasadena Fire Department's active roster until 1930, and has since been restored. BOB ALLEN.

This is San Diego's 1910 Seagrave aerial. The tractor's serial number is 3695 the ladder's is 4365. The motor was 90 hp and cooled by forced draught from a positive blower (which can be seen in the lower left), driven from the crankshaft. CHUCK PETERSON.

The first 65-foot aerial truck in Idaho was Boise's 1911 Seagrave. It was painted white and decorated with mermaids (appropriate for a make with the name Seagrave). This was a Model AC 80 (air-cooled, 80-hp motor), serial number 5832. The wheels were artillery-type of hickory wood, with steel channels for the solid rubber tires, single in front, dual in rear. Note brass hubcaps and chains — both front and rear — for traction on dirt streets. HANK GRIFFITHS.

142 • *Motorized Fire Apparatus of the West*

It's a long way down the ladder on Salt Lake City's 1913 Model WC 80 (water-cooled, 80-hp motor)Seagrave aerial. The aerial sits upon a fifth wheel under the ladder turntable, which is mounted on a rocker taking into account the unevenness of the street. SALT LAKE CITY FIRE DEPARTMENT.

seat to the side of the ladder. This reduced the vehicle's overall length.

The first Seagrave pumping engines used the Gorham multi-stage centrifugal pump (see Chapter Two). These were built as straight pumping units in the period 1912-1914. In spite of the Gorham's capabilities, Seagrave had also discovered a centrifugal pump used in England. This was marketed in the United States under the name "Manistee," which was superior to most pumps then sold in this country. Seagrave changed pumps, and the first Seagrave with a Manistee all-bronze, multi-stage centrifugal pump went to Santa Barbara, California, in 1914. Manistee pumps gave Seagrave an advantage over their competitors for much of the 1920s, and eventually the rest of the industry had to follow Seagrave's lead.

Seagrave remained a dominant force in the market for many years. It also outfitted commercial chassis. In 1963, it was acquired by FWD, and moved to Clintonville, Wisconsin, where it remains to this date.

A buckboard-style Seagrave which was purchased by the Napa (California) Fire Department in 1910. It was an AC 40 (air-cooled, 40-hp motor), two-tank chemical and hose car, serial number 4510. It's being followed by Napa's Seagrave-Gorham pumping engine, Model WC 144. ROBERT SAMS.

Motorized Fire Apparatus of the West • 143

This 1911 Seagrave AC 53 (air-cooled, 53-hp motor), serial number 5431, combination hose and chemical tank was used by Grand Junction, Colorado. The 60-gallon chemical tank was made of extra heavy Lake Superior hammered copper, and tested at 450 pounds of hydrostatic pressure. Note the revolving flood light on the dash and the two brass oil lamps. ROBERT SAMS.

This is a different type of Seagrave hood, and it is on a tractor purchased by Long Beach, California, in 1911 to pull a tillered city service truck. It is a Model AC 80 (air-cooled, 80-hp motor), serial number 7080. ROBERT SAMS.

The first motorized fire apparatus in Nevada was purchased by the Warren Engine Company of Carson City. It was a Seagrave Model WC 80 (water-cooled, 80-hp motor) combination hose and chemical wagon, serial number 10505, built in 1911. It has been restored and is displayed at the fire museum in Carson City. WAYNE SORENSEN.

Richmond, California, rebuilt their 1912 Seagrave Model AC 80 (air-cooled, 80-hp motor) combination chemical and hose car in 1922. The new motor was water-cooled and placed in front of the driver. PAUL DARRELL.

Santa Clara, California, used this 1913 Seagrave as a chemical and hose rig until it was converted to a 500-gpm pumper in 1942. The Santa Clara Fire Department's chief mechanic, Frank Menzel, developed a wiring system whereby the truck's starter was connected to the fire department's Gamewell Alarm system, so that when an alarm came in, the truck's engine would be started and warming up as the first volunteers arrived at the station. The rig has been restored, except for the hard rubber tires. SANTA CLARA FIRE DEPARTMENT.

This is a smaller Seagrave, a 1916 combination 375-gpm pumper and hose car, delivered to Douglas, Arizona. Its serial number, 15950. DICK ADELMAN.

Motorized Fire Apparatus of the West • 145

Santa Barbara, California, bought this 1919 Seagrave "sidesaddle" 65-foot aerial, serial number 22991. The driver had the greatest difficulty in seeing where the truck was going. The ladder was lifted by a full automatic spring hoist. Later, the unit was converted to a tractor-trailer aerial. ED GARDINER.

Seagrave also built two-wheel tractors to take the place of horses. These tractors were driven by double sidechains from Jack-shaft sprockets to sprockets on the wheel-driving shaft. This K4 tractor is attached to a 1900 American-LaFrance 700-gpm steamer in Seattle, Washington. SEATTLE FIRE DEPARTMENT.

In 1920, the first Seagrave with an improved "Manistee" centrifugal pump went to Santa Barbara, California. The rig's serial number was 24463, and the pump was rated at 900 gpm. SEAGRAVE.

Seagrave's styling changed in the early 1920s, as can be seen in this advertisement from *Fire Protection* magazine.

Seagrave also marketed a 400-gpm centrifugal pump for smaller communities. It was named "suburbanite." Here's a 1927 model, serial number 47662, purchased by the Carson (Nevada) Fire Department. It was powered by a six-cylinder Continental 70-hp engine. WAYNE SORENSEN.

The grille on this 1937 Seagrave is sometimes referred to as a "sweetheart" grille. This 65-foot aerial was purchased by the Pocatello (Idaho) Fire Department. It had a small booster tank, and its serial number was 87580. WAYNE SORENSEN.

Two of the manifold wagons that were built for Los Angeles by Seagrave in 1938. These wagons usually ran with duplex pumpers (see the American-LaFrance photo at the bottom of page 104). On this rig the hose bed is filled with 3½-inch hose, and the transverse beds carried 2½-inch hose for the 16 outlets. Note the fireboat-size monitor. On the left is Engine No. 17, serial number 88040; in the rear is Engine No. 3, serial number 88041. BOB ALLEN.

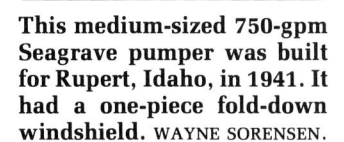

This medium-sized 750-gpm Seagrave pumper was built for Rupert, Idaho, in 1941. It had a one-piece fold-down windshield. WAYNE SORENSEN.

This type of bumper appeared on Seagraves in late 1948. This is Denver's (Colorado) Squad 2, a flood light unit, powered by a V-12 engine. Note the Mars warning light on top of the grille and the revolving Roto-Rays warning light system behind the seat. DICK ADELMAN.

148 • Motorized Fire Apparatus of the West

1932 Reo, outfitted by Seagrave for Walsenberg, Colorado. It had a 400-gpm pump, plus booster tank. FWD.

Port Angeles, Washington, purchased this 1936 series RD Chevrolet-Seagrave with a 500-gpm all-bronze centrifugal pump. This is a factory photo, taken when the rig was completed and ready for delivery. FWD.

Slightly beyond the time frame encompassed by our book is this 1962 Seagrave Model 900-B built for Las Vegas, Nevada. It shows Seagrave styling for the period 1952-1966. The grille appears to be smiling, and the siren is mounted in the hole above. The unit was powered by a Hall-Scott engine. The hood was specially-designed with extra ventilation slots for the warm Las Vegas weather. BILL FRIEDRICH.

Motorized Fire Apparatus of the West • 149

This 1941 Ford-Seagrave ran as Truck 2 in the Winslow (Arizona) Fire Department. It was equipped with a 500-gpm centrifugal pump and booster tank, and powered by Ford's popular V-8 engine. (Rigs on commercial chassis such as this all but replaced the market for the Seagrave Suburbanites. They were lower cost to begin with, and parts were available locally.) FWD.

This started out as a 1943 Seagrave 750-gpm quad for the United States Navy, but in 1962 it was rebuilt by Coast for Citrus Heights, California. A 65-foot Grove aerial ladder and compartments were added, as was Coast front end metalwork; also it was repowered with an International engine. WAYNE SORENSEN.

This is a 1958 Ford F-800-Seagrave with a 750-gpm pump built for the Payette (Idaho) Fire Department. It also had a 1,000-gallon water tank. JOHN SORENSEN.

150 • *Motorized Fire Apparatus of the West*

Berkeley was the first California city to buy a Stutz. Shown is a 1920 Model C 750-gpm pumper undergoing acceptance tests. HAROLD SCHAPELHOUMAN.

Stutz
Indianapolis, Indiana

Stutz fire apparatus was built by the Stutz Fire Engine Company from 1919 to 1928. The pumper was designed by A.C. Mecklenburg, who then approached Harry C. Stutz, who had just sold his interest in the Stutz Motor Car Company, maker of the famous "Bearcat." Stutz financed the new company and gave it his name, but didn't stay long, moving on to form yet another manufacturing company. An early Stutz pumper went on display at an annual meeting of the International Association of Fire Engineers, and in a pumping demonstration outdid all rivals. This gave the newcomer instant recognition and some orders were even received at that convention.

Hillsboro, Oregon, still owns this 1922 Stutz Model K-3 500-gpm pumper. The windshield and overhead ladder racks were added later. Note the red lens in the headlights. BOB ALLEN.

Denver had a fleet of five Stutz apparatus. This 1922 Model O, a 1,000-gpm pumper, ran as Engine 6. WAYNE SORENSEN COLLECTION.

Los Angeles must have been one of Stutz's best customers, buying 19 units in all! This is Engine 48, a Model JS combination chemical and hose car. DALE MAGEE.

This is Nampa, Idaho's 1924 Stutz Model O, a 1,200-gpm triple combination pumper, one of only nine that Stutz built. It is now being restored. NAMPA FIRE DEPARTMENT.

152 • *Motorized Fire Apparatus of the West*

Los Angeles used five of these 1924 Stutz Model TS ladder trucks. The long flat box on the top of the ladder bed is for the life net. The rear wheels are tillered. DALE MAGEE.

The firm lasted about a decade, during which it produced 235 pumpers, two quads, 18 chemical and hose cars, 44 city service trucks and one tiller-aerial. Stutz apparatus used Northern rotary pumps ranging from 350 gpm to 1,200 gpm. Some engines were made by Stutz. Stutz apparatus was popular in the West.

In the late 1920s, the firm was racked by internal feuding. Mecklenburg left and, with some former employees, formed a new firm. They acquired the rights to use the name "The New Stutz Fire Apparatus Company." This new organization lasted through the 1930s.

San Jose's (California) 1922 Stutz Model C 750-gpm triple combination pumper after extensive rebuilding in 1940. A Peterbilt grille covered a new Hall-Scott engine. Observe the new wheels, water tank on the hose bed and the San Jose Fire Department's color scheme. CHRIS CAVETTE.

Motorized Fire Apparatus of the West • 153

Above: **The United States Army post at the Presidio in San Francisco used this 1938 U.S.A. It was known as a Type 50 and carried a 750-gpm Hale centrifugal pump. These trucks were familiar sites on Army posts during World War II.** PAUL DARRELL.

Below: **This U.S.A. ended up in service at Rocky Mountain National Park in Colorado. A booster tank and hose reel have been added.** NATIONAL PARK SERVICE, ROCKY MOUNTAIN NATIONAL PARK.

154 • *Motorized Fire Apparatus of the West*

The Horseshoe Bend (Idaho) Fire Department acquired this U.S.A. after World War II. Examination of the metal tags showed that it had been originally built in the Fort Holabird motor transport shop as job order number 8000, dated June 15, 1938. The engine was Continental, 22 RF 1455, and the pump was a Hale, 2843, 750-gpm.
GAYLE SORENSEN.

U.S.A.
Fort Holabird, Maryland

In peacetime, the United States Army needs fire apparatus to protect its camps in the same way any community does. In the years between World War I and World War II, the Army was strapped for funds. Its Quartermaster Corps, at Fort Holabird, Maryland, decided to assemble their own fire apparatus. For vehicles they used "Liberty" trucks, a standardized World War I design that a number of United States commercial truck manufacturers had built. (There was a single design, and all parts were interchangeable.)

Most of these U.S.A. pumpers were built between 1937 and 1941, and they appeared at military posts throughout the country. (They were known as "type 50.") Most had Hale centrifugal 750-gpm pumps and Continental engines.

The Quartermaster Corps at Fort Holabird also built a smaller number of airport crash trucks for use by the United States Army Air Corps. These had double rear axles, booster tanks and 250-gpm pumps. They were powered by Lycoming engines.

After World War II, many of these U.S.A. engines became "surplus" and were acquired by fire departments throughout the country.

Fort Collins, Colorado, purchased this surplus U.S.A. 750-gpm pumper after World War II and painted it white.
WAYNE SORENSEN.

Motorized Fire Apparatus of the West • 155

Above: This U.S.A. six-wheel crash truck was used at an air base at Oakland, California. The device on the running board was a hopper for mixing foam concentrate with water. The rig was powered by a Lycoming A.F.D. 130-hp engine. PAUL DARRELL.

Below: The Quartermaster Corps developed the Type 50 apparatus in 1932, with six wheels (the front non-driving) for use around United States Army Air Corps installations. This 1946 model was later acquired by the Danville (California) Fire Department and built into a 1,000-gallon tanker. WAYNE SORENSEN.

This is one of the first Ward LaFrance rigs used in California. It's a 1948 750-gpm pumper, delivered to Rio Vista. The post-mounted siren on the cowl was standard, the rear fender skirts optional. WAYNE SORENSEN.

Ward LaFrance Corporation
Elmira, New York

In 1918, this firm was founded by Ward LaFrance, a member of the family associated with the builders of American-LaFrance fire apparatus. There was no commercial relationship between the two firms, however. Ward LaFrance built "assembled" trucks, meaning that they purchased all their components from outside suppliers and then assembled to complete products. In addition to fire apparatus, the firm built trucks for a number of other government and commercial applications.

In 1931, Ward LaFrance built their first engine. Soon, fire apparatus became the most important element of their business. They marketed both custom apparatus and apparatus on commercial chassis. They also produced a number of airport crash trucks. Their most noticeable contribution to the fire apparatus field was in terms of color. Ward LaFrance was the first to replace "fire engine red" with the bilious yellow-green, which is more visible at night.

The firm is no longer in business.

This 1944 Ward LaFrance six-ton 6x6 crash truck was used at the San Francisco airport. It was powered by a Continental Model 22R six-cylinder 145-hp engine. A second engine (a Continental R602) drove the pumping equipment and the high-pressure foam system. Hand-operated nozzles on top of the cab were aimed by hand. WAYNE SORENSEN.

Castro Valley, California, bought this 1948 Ward LaFrance with a 750-gpm two-stage Hale centrifugal pump. It also had a 300-gallon water tank. The Castro Valley Fire Department took delivery at Elmira, New York, and several of their firemen drove it across the country. WAYNE SORENSEN.

This 1953 Ward LaFrance Model CW 750-gpm pumper was built as part of a United States Air Force order for 662 units. The rigs were intended mainly to fight structural fires. This one was purchased from the military by Anchorage, Alaska. It carried a Waterous 750-gpm pump and both a 175-gallon water tank and a 40-gallon foam tank. It was powered by a Continental Model R6602, 240-hp engine. WAYNE SORENSEN.

Ward LaFrance also outfitted commercial chassis, such as this 1941 Ford with a 500-gpm Hale pump and small water tank for the military. Later, it was used by the fire department at Diamond Springs, California. WAYNE SORENSEN.

158 • *Motorized Fire Apparatus of the West*

Emeryville, California, bought this 1912 Webb pumping engine and hose car built on a Thomas touring car chassis and equipped with a 600-gpm rotary pump. EMERYVILLE FIRE DEPARTMENT.

The Webb Company
Allentown, Pennsylvania

Al C. Webb, a race car driver from Joplin, Missouri, built a chemical car on a Buick chassis for the Joplin Fire Department in 1907. He also converted a Thomas Flyer chassis to a pumper. Early in 1908, he was asked by the Lansing, Michigan, fire chief to build a hose and pumping engine on an Oldsmobile chassis, which he did. Webb then organized the Webb Motor Fire Apparatus Company, and located in Vincennes, Indiana. The firm prospered. It was reorganized and expanded, then relocated, for a short time, to St. Louis where it received some local backing.

In 1912, a large block of its stock was sold to John W. Mack, and the firm moved to Allentown, Pennsylvania. There, in addition to fire apparatus, the company built police patrol wagons and ambulances. They apparently dropped out of business about the time of World War I.

The Caldwell (Idaho) Fire Department drilling in the 1920s at city hall with their 1914 Webb 600-gpm pumper. CALDWELL FIRE DEPARTMENT.

Motorized Fire Apparatus of the West • 159

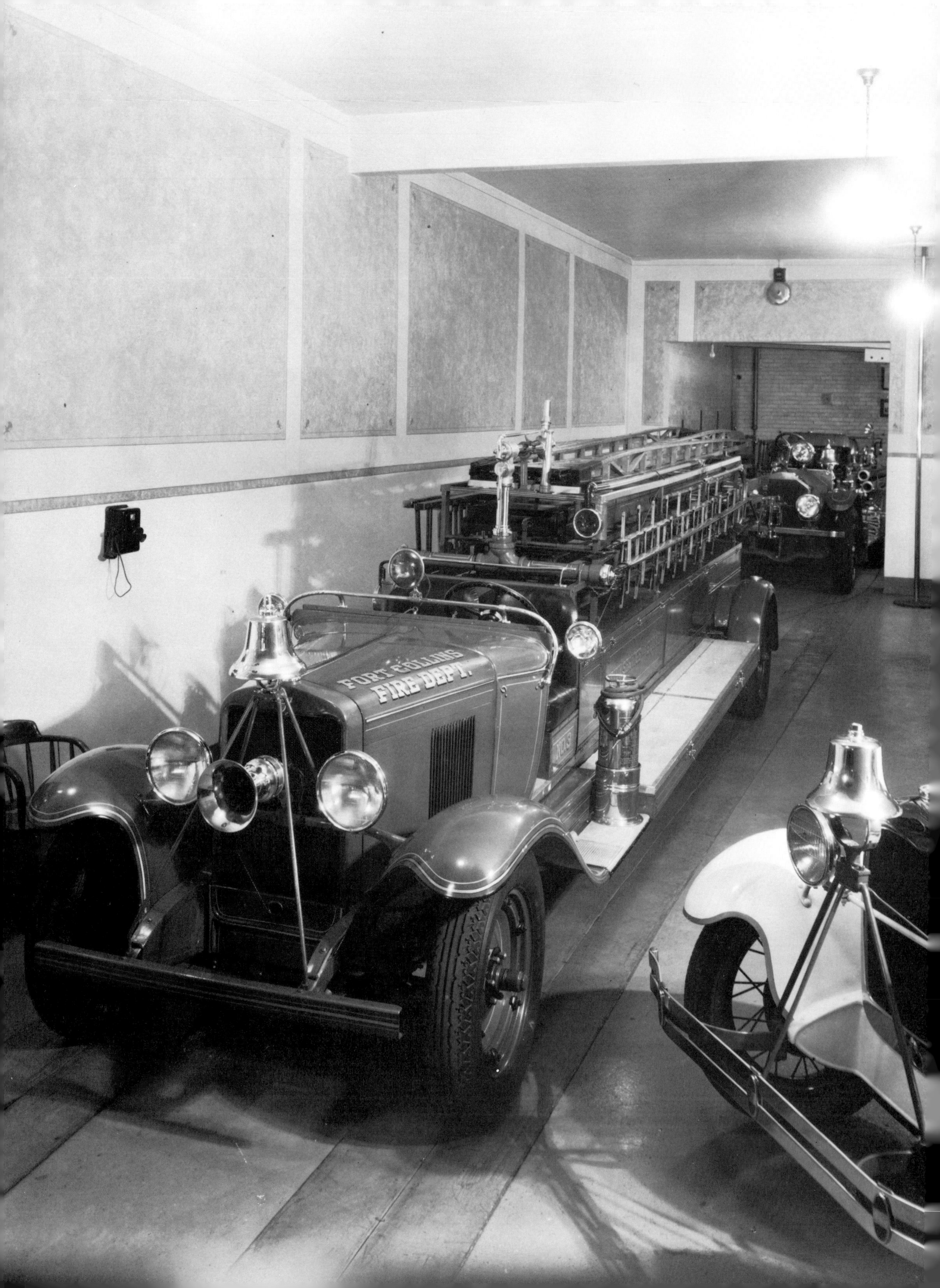
FORT COLLINS
FIRE DEPT.

4

More Apparatus on Commercial Chassis

Both Chapters Two and Three contain many examples of commercial chassis outfitted with fire apparatus made by known manufacturers. This last chapter has more examples of commercial chassis used to carry fire fighting bodies. For the most part, we don't know who built the body. No doubt, in some instances, the body builder was one of the firms already discussed, but at this late date, complete documentation on the vehicle no longer exists.

However, we believe that body builders of much of the apparatus pictured in this chapter were small shops, which did little more than attach a purchased pump to a chassis, add a hose bed and weld on brackets to hold ladders and other tools. A few of the rigs pictured look that basic.

This chapter is alphabetized by make of chassis manufacturer (if known). When we inserted information about the chassis builder, our usual source was: *The Complete Encyclopedia of Commercial Vehicles*, G. N. Georgano, editor, Iola, Wisconsin: Krause Publications, 1979.

Above: Contra Costa County's Consolidated Fire District, headquartered in Pleasant Hill, California, used this 1942 Autocar wrecker, outfitted with Holmes twin booms. BOB ALLEN.
Preceding Page: Fort Collins, Colorado, lengthened the frame on this 1930 Chevrolet to produce a handsome city service truck. FORT COLLINS PUBLIC LIBRARY.

Motorized Fire Apparatus of the West • 161

Biederman trucks were made in Cincinnati, Ohio, from 1920 until 1955. This Biederman Model tff 010 crash truck was originally built for the United States Air Force, and later acquired by the Eleverta (California) Fire Protection District. This massive rig carried 2,000 gallons of water and has a 2,000-gpm pump. WAYNE SORENSEN.

Colusa, California, used this 1915 Brockway, which had been outfitted with American-LaFrance chemical equipment. It has been restored. WAYNE SORENSEN.

Brockways were manufactured in Cortland, New York, until only a few years ago. They are rare in the West. Here's a 1918 Brockway chemical and hose car, purchased by the United States Army and stationed at the Presidio of Monterey, California. The body is probably an American-LaFrance. MONTEREY FIRE DEPARTMENT.

162 • *Motorized Fire Apparatus of the West*

Buick is a well-known maker of autos. Early in this century, however, they also produced light trucks. Here is a 1914 Buick Model D-4 hose wagon, used by the Auburn (California) Fire Department. It has been restored. WAYNE SORENSEN.

The volunteer fire fighters of Mission San Jose (now a part of Fremont), California, used this 1925 Buick Master 6 roadster chassis to carry two 30-gallon chemical tanks to make a chemical car. Note the floodlight. WAYNE SORENSEN.

This 1933 Buick auto was converted to a tractor to pull a one-time horse-drawn tillered city service truck. It first saw service in Richmond, California; then it was at Walnut Creek, California. PAUL DARRELL.

Motorized Fire Apparatus of the West • 163

The Oakland (California) Fire Department's shops selected this 1914 Cadillac touring car to build a tractor for Truck 5. One reason for this choice was the Cadillac's dependable Delco system of electric starting, lighting and ignition. The fifth wheel was installed over the rear axle. CLANCY CRUM.

Pleasanton, California, purchased this 1914 Cadillac combination chemical and hose car. It had right-hand steering. Note the windshield, unusual for fire apparatus of this vintage. ROBERT SAMS.

Nampa, Idaho, converted this 1916 Type 33 Cadillac auto into a chemical car, with two 60-gallon tanks. The motor was an early V-8. NAMPA FIRE DEPARTMENT.

164 • Motorized Fire Apparatus of the West

Atwater, California, converted this 1918 Cadillac auto into a 350-gpm pumper by adding a front-mount pump and hose body. WAYNE SORENSEN.

When Ripon, California, decided to restore its 1925 Chevrolet pumper, it sold advertising space on the truck's sides to help raise funds for the restoration. This rig has a 250-gpm front-mount pump and a 300-gallon tank. WAYNE SORENSEN.

Chatsworth, California, bought this 1928 Chevrolet that has a booster tank and hose body. GEORGE "SMOKY" BASS.

Motorized Fire Apparatus of the West • 165

Above: This 1929 Chevrolet chassis was purchased and outfitted by volunteers at Virginia City, Nevada. It has a 100-gpm pump, a 250-gallon tank, and a hose reel (mounted in place of the passenger's seat). The rig is seen at musters. WAYNE SORENSEN.

Below: Rio Nido, California, ran this 1930 Chevrolet-American-LaFrance Type 36 combination booster car, with equipment installed by the American-LaFrance San Francisco service center. WAYNE SORENSEN.

166 • *Motorized Fire Apparatus of the West*

The Live Oak Fire District near Santa Cruz, California, took a 1934 PA Chevrolet, which had been used as an oil truck, and built their first fire engine. It was equipped with a separate pumping engine, 250-gpm pump, 87-gpm pressure pump and 400-gallon tank. Later, it was sold to Davenport, California. Now it is in private ownership. WAYNE SORENSEN.

This is a 1936 Chevrolet, used by the United States Forest Service in Montana. From the rear you can see much of its equipment and detail: railing for fire fighters also protects the hose reels; six shovels are in the rack above the water tank; a tarpaulin covers the rack for the canvas hose; flashlights, nozzles and fire extinguishers are on the racks; and there is a crew seat behind the cab. FOREST SERVICE, U.S.D.A.

This was originally a 4x4 military crash truck on a 1942 Chevrolet chassis. After World War II, Van Pelt rebuilt it with a 250-gpm front-mount pump and a 450-gallon water tank for Pollock Pines, California. WAYNE SORENSEN.

Motorized Fire Apparatus of the West • 167

Above: Burlingame, California, built this hose wagon on a 1942 Chevrolet COE (cab-over-engine). The rig also has a 500-gpm pump and a Gorter high-pressure turret. WAYNE SORENSEN.

Below: Coleman, of Littleton, Colorado, has built all-wheel-drive trucks and conversion kits since 1926. Here is the Littleton Fire Department's 1925 Coleman pumper. LITTLETON HISTORICAL SOCIETY.

168 • *Motorized Fire Apparatus of the West*

Corbitt trucks were built in Henderson, North Carolina, until 1958. This 1942 Corbitt is a military blimp tender, used by the United States Navy at Moffitt Field, California, during World War II. It carries fire fighting and rescue equipment. Boom is used to guide blimp along the ground. Note protective clothing. GEORGE P. HANLEY.

The Couple-Gear tractor was built in Grand Rapids, Michigan, and powered by electric motors driving its wheels. This 1912 Couple-Gear was linked to an Ahrens-Fox steam pumper, which had been built for that specific purpose (i.e. it was not a one-time horse-drawn unit); it was sold to Denver, Colorado, in 1912. AHRENS-FOX.

Crosleys were built in Indiana from 1940 until 1952, and today would be considered as subcompact-size vehicles. Here you see an industrial fire truck mounted on a 1950 Crosley chassis. Equipment includes a Bean-Rool high-pressure pump, a 60-gallon water tank, 200 feet of high-pressure hose and two five-gallon Indian packs. This unit served at the Aerojet General Machine Shops at Arden, California. The rig is now privately owned. WAYNE SORENSEN.

Motorized Fire Apparatus of the West • 169

Above: DeMartini trucks were built in San Francisco from about the time of World War I until the early 1930s. (One of this book's co-authors once interviewed the firm's founder who said he didn't like to be involved in the sale of fire apparatus because of the graft prevalent at that time.) The 1924 photo shows a DeMartini chemical and hose car used by the Mill Valley (California) Fire Department. LAURENCE F. JONSON.

Below: The Diamond-T Company of Chicago built the chassis for this 1939 COE (cab-over-engine) rescue truck used by Pasadena, California. COEs had shorter turning radii, took up less room in the station and gave the driver a better view. DALE MAGEE.

170 • *Motorized Fire Apparatus of the West*

A Diamond-T chassis was built for this 1940 city service truck, used by the Hillsborough (California) Fire Department. The ladders are fully enclosed. BOB ALLEN.

Rear view of Rio Lina, California's restored 1922 Dodge chemical car, which carries equipment installed by the San Francisco American-LaFrance service center. According to truck registrations, as of December 31, 1921, the Dodge Brothers' truck was the fourth most popular make in the United States. WAYNE SORENSEN.

This 1929 Dodge flatbed was outfitted with a 600-gallon water tank and a 250-gpm Barton front-mount pump, and was used by the Elmira (California) Fire Department. WAYNE SORENSEN.

Motorized Fire Apparatus of the West • 171

Above: The Columbia (California) Volunteer Fire Department still owns this 1930 Dodge, with a front-mount 250-gpm pump. BOB ALLEN.

Below: Proctor-Keefe, of Detroit, is best known as the manufacturer of a full line of commercial truck bodies. They did outfit some fire apparatus, such as this 1934 Dodge Series K 1½-ton chassis for Gallup, New Mexico. NATIONAL AUTOMOTIVE HISTORY COLLECTION, DETROIT PUBLIC LIBRARY.

172 • Motorized Fire Apparatus of the West

Quincy, California, ran this 1934 Dodge K with a 350-gpm front-mounted pump. PAUL DARRELL.

This 1940 Dodge Y, COE (cab-over-engine) was used as a light wagon by the Brisbane (California) Fire Department. WAYNE SORENSEN.

After World War II, Dodge marketed its "Power Wagon," which contained many features found on light United States Army trucks. Pollock Pines, California, used this 1954 Dodge Power Wagon as a squad truck. The snowplow was probably used to clear the driveway between the station's door and the street. WAYNE SORENSEN.

Motorized Fire Apparatus of the West • 173

Above: Dorris trucks were made in St. Louis from about 1912 to 1926. This 1926 Dorris hose wagon was used by the Madison (Missouri) Fire Department. CHUCK RHOADS.

Below: Fageol trucks were built in Oakland, California, from 1916 until 1939. They could be recognized by a row of shark-fin vents along the top of the hood. The Citrus Heights (California) Fire Department modified this 1925 Fageol gasoline tank truck into a fire fighting rig by adding a 150-gpm pump. The tank held 800 gallons. ED GARDINER.

The fire department shops in Spokane, Washington, used this 1936 Fageol chassis to build a city service truck, which saw service for 26 years. It was powered by a Hall-Scott engine. DICK ADELMAN.

Federal trucks were built in Detroit from 1910 until 1954. This 1914 Federal chemical and hose car was used in Littleton, Colorado. Note chain drive. BOB ALLEN.

Nampa, Idaho, used a 1916 Federal chassis to carry a chemical and hose car with an overhead ladder rack. NAMPA FIRE DEPARTMENT.

Motorized Fire Apparatus of the West • 175

Above: The Fresno (California) Fire Department shops used a 1950 Federal chassis to build a hose wagon with a large hose monitor. Siamesed 2½-inch inlets on the side feed the monitor. The truck was powered by a six-cylinder Hercules engine. BOB ALLEN.

Below: The Flxible Company of Loudonville, Ohio, built buses. This 1954 Flxible bus was used by the Bend (Oregon) Fire Department as a city service truck until 1980. This unit was built by incorporating ladders, rack and other parts from the Bend Fire Department's 1923 GMC-American-LaFrance city service truck. One reason for using the enclosed body was protection from the rain. WAYNE SORENSEN.

176 • *Motorized Fire Apparatus of the West*

This 1906 Ford N was used by Chief Frank "Buddy" Dowell of the Portland (Oregon) Fire Department. The Ford N auto helped Henry Ford become an established manufacturer. ALFRED C. JONES.

This Ford Model K was used by the Deer Lodge (Montana) Fire Department. The tonneau has been replaced with a hose body, and it's pulling a two-wheeled chemical cart. POWELL MUSEUM, DEER LODGE, MONTANA.

The Ford Model T was introduced in 1909, but it was not until 1917 that Ford introduced a truck chassis, the Model TT. However, between these two dates, a number of manufacturers developed methods of lengthening and straightening the frame. Here's one such example, used in Los Angeles County, where a chemical and hose body, purchased from American-LaFrance, was mounted on a lengthened 1914 Ford chassis. RICHARD JAMES.

Magalia, California, ran this 1914 Ford booster truck, which carried a water tank and small pump mounted on the rear. A "carry-all" is mounted in the running board to carry extra equipment. WAYNE SORENSEN.

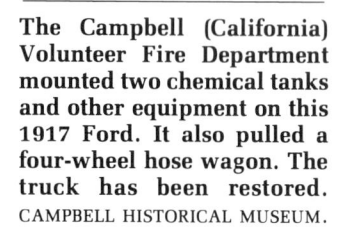

The Campbell (California) Volunteer Fire Department mounted two chemical tanks and other equipment on this 1917 Ford. It also pulled a four-wheel hose wagon. The truck has been restored. CAMPBELL HISTORICAL MUSEUM.

Volunteer fire fighters at Willow Oak Park (California) used this 1919 Ford TT flatbed to build their fire engine. It carries a tank and a John Bean pump, driven by a separate motor. Fords were much more common in the fire fighting service during this era than space in this book implies. They were low-cost, easy to drive and maintain and a dealer and parts were close at hand. (As of December 31, 1921, slightly over one-half of the trucks registered in the United States were Fords.) WAYNE SORENSEN.

178 • Motorized Fire Apparatus of the West

Above: This 1920 Ford TT was built by the Orland (California) Volunteer Fire Department. They fashioned the wooden tank from an old orchard spray rig. A small gasoline pump is at the rear. WAYNE SORENSEN.

Below: Los Angeles County used Ford Model T autos to pull trailers with fire fighting equipment. We can see just the end of one such auto on the right. The dirigible is being used for observation purposes. Photo was taken in 1921. FOREST SERVICE, U.S.D.A.

Motorized Fire Apparatus of the West • 179

Above: **In the last years of its production, the Model T took on a slightly more sturdy look. This is a 1926 Ford with a wooden hose body, used by the Larkspur (California) Fire Department. Note the chain sprocket. The truck still runs in parades.** WAYNE SORENSEN.

Below: **The Los Altos (California) Volunteer Fire Department still owns this 1928 Ford AA pumper. It has a 300-gallon water tank and a 250-gpm pump.** WAYNE SORENSEN.

180 • Motorized Fire Apparatus of the West

This 1930 Ford AA city service truck was used by the Mountain View (California) Fire Department until 1963. It had been built by the Redwine Motor Company of Mountain View. MOUNTAIN VIEW FIRE DEPARTMENT.

This 1930 Ford Model A pickup was employed by the San Francisco Fire Department as a "valve" wagon. It carried a hydraulic spindel arm, used to turn off high-pressure hydrants. Note opening in ground; it's a high-pressure valve with its lid off. BOB ALLEN.

1932 Ford BB was used by the Mills (California) Fire District, now part of the California Rancho Cordova Fire Protection District near Sacramento. A separate Hercules gasoline engine powers a 250-gpm pump. The rig is still used in parades. WAYNE SORENSEN.

Motorized Fire Apparatus of the West • 181

Above: The California state fire marshall's office used this 1937 half-ton Ford V-8 pickup to pull a FABCO-built trailer, which carried a small gasoline pump and hose. FABCO.

Below: In 1938, the Fresno (California) Fire Department's shops built four rigs, known as "Fresno Specials," with 1,000-gpm pumps. Here's one of them. (In the early days of motorized apparatus, the Fresno Fire Department entered into an agreement with the local blacksmith shop of Larson and Krog to use the blacksmith's shop for building apparatus. Firemen were released from their fire fighting duties to work in the shop, and given the title "Fire Department Repairmen.") WAYNE SORENSEN.

Above: This "Fresno Special," rebuilt in 1950s by Van Pelt with Hall-Scott power and Peterbilt grille and hood. BOB ALLEN.
Below: This 1924 GMC chemical and hose car was operated by the San Jose (California) Fire Department. It has a 35-gallon chemical tank. WAYNE SORENSEN COLLECTION.

Motorized Fire Apparatus of the West • 183

In several instances through this book, it's been mentioned that apparatus was rebuilt. This and the following picture illustrate such changes. Here, we see a 1927 GMC two-tank chemical and hose car, purchased by the Piedmont (California) Fire Department. PIEDMONT FIRE DEPARTMENT.

In 1935, Piedmont had the rig shown in preceding photo rebuilt. The chemical tanks were removed, and a booster tank and hose added. A windshield was added, as was a front-mounted 350-gpm pump. Headlights were moved from the cowl to the front. PIEDMONT FIRE DEPARTMENT.

The Oakland (California) Fire Department shops built this light wagon on a 1937 GMC T chassis. It carried eight large floodlights, portable lights and cable and a 10-kw generator. WAYNE SORENSEN.

184 • Motorized Fire Apparatus of the West

This 1940 GMC-AFX chassis was used by the Glenn Ellen (California) Fire Department for a pumper. A second engine drives the pump. BOB ALLEN.

This lightly-equipped 1941 GMC ran as a chief's car in Butte, Montana. The body work was done by Howard-Cooper, and the rig carried a small Oberdorfer pump. HOWARD-COOPER.

In the 1920s, there was a close relationship between Graham Brothers and Dodge trucks. This is a 1924 Graham Brothers which had been used as a taxicab chassis until 1940. Then it was converted into an industrial fire rig for use at a Lodi, California, winery. It has a 250-gpm pump and a 150-gallon water tank. It has been restored by the Woodbridge (California) Firemen's Association. WAYNE SORENSEN.

Motorized Fire Apparatus of the West • 185

Above: **This attractive 1924 Graham Brothers-American-LaFrance Type 36 chemical and hose wagon was used by the Tulare (California) Fire Department. The fire fighting equipment had been added by the American-LaFrance San Francisco service center. Notice the Budd-Michelin steel disc wheels.** AMERICAN-LAFRANCE SERVICE CENTER.

Below: **This 1928 Graham Brothers one-ton chassis was used by the Berkeley Fire Department to build a light and rescue unit.** HAROLD SCHAPELHOUMAN.

186 • Motorized Fire Apparatus of the West

In front of Bakersfield, California's Fire Station 2 is a 1928 Graham Brothers that pulled a one-time horse-drawn city service truck. BOB ALLEN.

Gramm trucks were built in Lima, Ohio. This 1909 model was used as a chemical and hose car in Colorado Springs, Colorado. A.L. HANSEN MANUFACTURING COMPANY.

The United States Forest Service used the Harley-Davidson motorcycle in the Gila National Forest of New Mexico. FOREST SERVICE, U.S.D.A.

Motorized Fire Apparatus of the West • 187

Above: Hewitt-Ludlow trucks were built in San Francisco from about 1912 until 1926. A Hewitt-Ludlow chemical and hose car was used by the Sausalito (California) Fire Department from about 1925 until 1935. Here's a 1910 Seagrave chemical and hose car belonging to the Piedmont (California) Fire Department after it was rebuilt in the Hewitt-Ludlow shops following an accident. During the rebuild, it was given a Hewitt-Ludlow radiator and hood and a more powerful engine, needed for the hills of Piedmont. Much of the rebuilding work was done in the Ralston Iron Works of San Francisco. PIEDMONT FIRE DEPARTMENT.

Below: Hudson built passenger cars and, in some years, light trucks. This 1926 Hudson Super Six auto had its frame lengthened and straightened and the rear part of the passenger body removed by the Homestead (California) Volunteer Fire Department in order to make this rig. WAYNE SORENSEN.

188 • Motorized Fire Apparatus of the West

Hug trucks were made in Highland, Illinois, between 1922 and 1942. The San Antonio (Texas) Fire Department shops used a 1937 Hug Model 23AS chassis and installed a pump and other equipment to build this rig. CHUCK RHOADS.

A rear view of the same rig.
CHUCK RHOADS.

International Harvester has been building trucks since 1907, and in 1986 changed its corporate name to Navistar International. In 1923, the Lebanon (Oregon) Fire Department bought this 1917 International, which was a Portland milk truck at the time, and converted it into a hose wagon. It was in service until the late 1950s, and is still used in parades. Note the **gong.** WAYNE SORENSEN.

Motorized Fire Apparatus of the West • 189

Above: For a short period of time International sold its own line of fire apparatus. Here are two 1924 Internationals, bought by the Petaluma (California) Fire Department. This picture originally appeared in a 1926 edition of *The Motor Truck*. The caption said: "Petaluma, California, a thriving city of 7,000 inhabitants, which is noted for the hundreds of white leghorn chicken ranches in its vicinity, is winning fame also in the coast region for its superior fire fighting apparatus. . . . The city owns six trucks, two of the latest editions being the two Model 33, 1½-ton Internationals shown herewith. Five regular firemen are on duty by day and eight by night. In addition there are 13 call or part-pay firemen, all of whom have automobiles so that they can quickly respond to alarms. Bells and sirens, operation of which is controlled from one of the firehouses, are installed in all main thoroughfares to warn traffic in time of fire. One of the new Internationals, designated as No. 4 truck is a combined hose and chemical truck; it carries a 40-gallon chemical tank, 1,200 feet of 2½-inch hose, 200 feet of 1½-inch hose, two ladders, two three-gallon Pyrene extinguishers, two tubes of dry chemicals for chimney fires, one acid can, a full complement of shut-off nozzles, etc. The other is known as truck 5 and is used as a hook and ladder unit. It carries . . . a 50-foot extension ladder, life net, powerful searchlight, electric siren, etc." NAVISTAR INTERNATIONAL.

Below: A 1937 International 500-gpm pumper, of the type used in the West. INTERNATIONAL HARVESTER.

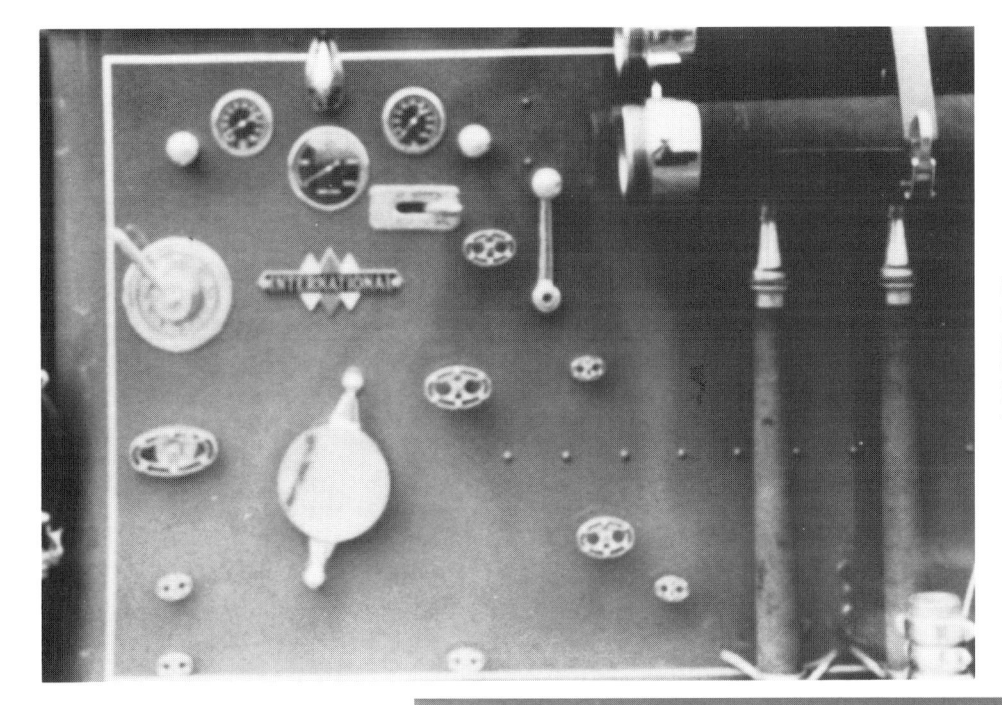

Close-up of pump panel of engine shown at bottom of prior page. Note International three-diamond emblem. INTERNATIONAL HARVESTER.

The Alameda, California, Naval Airbase World War II-vintage crash truck, built on an International M-5-6 series chassis. It had a high-pressure fog system. The fog was applied by the hand-operated turret nozzle mounted on the top of the cab. PAUL DARRELL.

Elko, Nevada, bought this 1959 International with Pitman snorkel equipment. It also had a 750 gpm pump and a 300-gallon water tank.

Motorized Fire Apparatus of the West • 191

Above: Jefferys were built in Kenosha, Wisconsin, during the World War I era. (Some "Ramblers," manufactured by the company as autos, are pictured later.) Here's the Oakland (California) Fire Department's Hose Company 4 responding with a large crew on their 1915 Jeffery. CLANCY CRUM.

Below: Kelly-Springfield trucks were built in Springfield, Ohio, from 1910 until 1929. Seattle, Washington, ran this 1913 Kelly-Springfield hose wagon. SEATTLE FIRE DEPARTMENT.

192 • *Motorized Fire Apparatus of the West*

Above: This is identified as a 1925 Kelly (built by Kelly-Springfield) serving as a chemical and hose car at Westminster, Colorado. BOB ALLEN.

Below: Kenworth has been building trucks in Seattle since 1923. They have assembled some apparatus in their own shops, and also serve as supplier of chassis to apparatus builders in the area. This is the first Kenworth fire engine chassis assembled, a 1932 500-gpm pumper sold to Sumner, Washington. The vehicle is still used in parades. KENWORTH.

Motorized Fire Apparatus of the West • 193

This Pirsch 85-foot hydro-mechanical ladder was pulled by a 1939 Kenworth tractor, which replaced the original White tractor that had been destroyed in an accident. The rig was owned by the Spokane (Washington) Fire Department. KENWORTH.

United, a Southern California aircraft manufacturer, built this quad body on a 1940 Kenworth chassis for the Los Angeles Fire Department. The rig had a 500-gpm pump. Firemen referred to the design as "Coca-Cola truck." BOB ALLEN.

United built the rakish body on this 1940 Kenworth 1,000-gpm pumper for Beverly Hills, California. After entering the cab, one had to walk forward to the driver's seat. BOB ALLEN.

194 • Motorized Fire Apparatus of the West

CHASSIS RECORD		FINAL		KENWORTH MOTOR TRUCK CORPORATION	
Wheelbase:	195"	Speed:	58	30 Cab Type: Cowl	Serial: 8577
BC:	160"	CA:	101-1/2"	32 Gas tank:	8573
Chassis weight:	10405	Front:	5530	34 Dash: With Cowl	
Gross:	20700	Front 7000	Rear 14700	35 Fenders: 8578	
1 -12 Axle, front	Timk	No.	35000-H	36 Running Board: with body	
Serial:	X-126	Axle shaft No.	C-965	37 Hood: Special	
2 -57 Axle, rear	Timk	No.	58300-H	38 Radiator: 38-6974	
Serial:	X-2	Ratio:	5-1/8	Shutter:	Core: Young 20472
Drive Sprocket:		Jackshaft:		39 Engine: Ha-Scott Model: 177	Pistons: Alum
Chain:		Dead Axle:		Serial: 560082	Water Pump: Ha-Scott
3 Hubs, front	Timk	No.	333-X-180	Oil filter: Hall-Scott	Air cleaner: Screen-Air Maze
4 Hubs, rear	Timk	No.	333-S-149	41 Fan Belt:	Their:
4 Hubs, Dead Axle		No.		42 Governor:	at R.P.M.
5 Studs, front: 32309-1G, 12247-8 Rr.			No. 6	43 Carburetor: Zenith	Model: 458 1570
5 Wheels:	Budd	No.	30398	45 Fuel Pump: 3 Bk. Auto Pulse & Cam	
6 Rims: Firest	Type: K			46 Electric units: D-K	Coil: 535-A
7 Tires: Gd Rch	Ply: 10	Size: 9.00-20		Generator: 405 OK-16	Watt capacity: 600
Dual Rear		Tread: Hiway & Tractor		Distributor: 4145 #4286	Magneto: Vert.Scintilla
8 Tire carrier:				Starter: 651 9L1	Voltage regulator: 5526 OL18
9 Springs, front:	Benz	No.	8573	Voltage: 12 Battery: Glb No: 64	No. 4
10 Springs, rear:	Benz	No.	8572	47 Lamps: Guide	Their: 675-N&P
				48 Instruments St.Vrn.	Cable: 4949 75"
11 Brakes, service: Hydraulic		On 4 Whs.		Drive gear: 72170	Driven gear: 70516
Size:	Type:			Adapter: 21941 Tach. No. Spec.	Cable: 4949 128
Serial:	Mast. Cyl: 1-3/4	Lever		51 Steering: Ross	Series: 720
Trailer coup:	Front Wh. Cyl: 1-1/2	Rear 1-3/4		Ident No. 730702	
12 Brakes, emerg. Truotop Model: Single				52 Drag link size: 1-1/2" X 36"	
on Crossmember - 16" Dia.				Colors:	
13 Booster: B-K	Cyl. JP-10	Valve:	RXD		
14 Brake drums, front: 3219-E-472					
Rear: 3219-A-573				Turn Signal:	Spots: On Rr. Body
15 Transmission, unit: Brown Lipe				Horns: 1880227	
Model: 6440	Serial: T-314773			Remarks: Carrier A30-N-92	
Clutch: B-L	Type: 13" - 2 plate			Bosch electric swipes	
16 Transmission, M.F.				10" S & M Searchlight -R.H.S.	
Model:	Serial:			B & M Siren on fender-L.H.S.	
17 Muffler SK-1029				2 Ruby Red cowl lights	
18 Drive shaft, front: Sol	Tube: 3"x.095"			Tachometer on pump panel only	
Joints: 7709-SF 1600 Ser.	Length: 53-1/4" CS			Insulate exhaust	
19 Drive shaft, center	Tube:			Battery boxes SK-1080	
Joints:	Length:			Body #8580	
Center brng. SKF:				Booster tank 8569	
SKF box:	Required:			Hose reel 6936	
21 Drive shaft, rear: Sol	Tube: 3" x .095"			Pipe layout 8584	
Joints: 7709-SF 1600 Ser.	Length: 83-1/2"CS			ZIED Hale pump serial #8433	
Drive shaft, between axle set				H- Rotary booster pump	
Joints:	Their Pt. No.			Extra tire & wheel	
22 Frame: S2-3027-D				Two siren control buttons-1 Ea. Side	
23 Frame reinforcement length:				Equipment as per specifications	
Outsert:	Insert:			Frame layout #L-8535	
25 Bumper: S5-8570					
27 Tow hitch, rear:	V brace:			Chassis Wt. Comp. with body but no	
U bolts:	Trailer hook:			equip. except gen.set. siren.searchlights	
28 Tow hooks, front: 8570	Rear: 30-Pmhole			10 Gal. gas -Tot. 14200. Fr.-6750. Rr.-7450	

Date 2-17-41	EL SEGUNDO FIRE DEPARTMENT	Model 731
By J.F		Serial 50956

The El Segundo Kenworth's original chassis record is reproduced here; note the additional equipment for a fire engine listed in the lower right. The completed apparatus is pictured on the following page. KENWORTH.

Motorized Fire Apparatus of the West • 195

Above: El Segundo, California, purchased this 1941 Kenworth. KENWORTH.
Below: This 1943 Kenworth tractor, with a 220 h.p. Hall-Scott motor, pulled a 1923 85-foot Seagrave aerial for the Los Angeles Fire Department. The aerial ladder trailer was one of three modernized by the United Fire Equipment Company of Los Angeles. BOB ALLEN.

Kissels were a well-known make of cars and trucks, built in Hartford, Wisconsin, up until 1931, and, for a while, the company may have marketed the fire apparatus they assembled. This is a 1914 Kissel chemical and hose car used by Salida, Colorado. FIREMEN'S HERALD.

This Kissel "Gold Bug" speedster was used by the fire chief in Butte, Montana. This is a Kissel Kar, a name the firm used for a period of time. Note the chemical tank; the chief would not have to stand by helplessly if he arrived first. FIREMEN'S HERALD.

The first fuel truck used by the San Francisco Fire Department was this 1918 Kissel. Used for refueling fire apparatus, it carried 500 gallons of gasoline and 100 gallons of oil, which says something about the engines of that time. BOB ALLEN.

Motorized Fire Apparatus of the West • 197

Above: Kleiber was a relatively small San Francisco firm that built trucks from 1914 until 1937. In 1926, the San Francisco Fire Department bought two Kleiber tractors to pull city service trucks. They proved to be too slow for ladder companies, so they were reassigned to pull water towers. This one is attached to a Gorter water tower; it's currently assigned to the San Francisco Fire Department Museum. BOB ALLEN.

Below: This is one of two 1928 Kleiber light wagons used by the San Francisco Fire Department. The bodies were built by the department's shops and included a 10-kw generator, five large flood lamps and eight portable lights. One of the trucks remains, and is currently owned by the Kleiber family. JOHN GRAHAM.

198 • Motorized Fire Apparatus of the West

The chassis of a 1924 Lincoln was used to build this light wagon for the Salinas (California) Fire Department. The chassis was extended two feet, and it has Cadillac front brakes, a 1935 International cab, a four-cylinder International engine and a 32-volt generator. The rig is now in private ownership. TOM MALLERY.

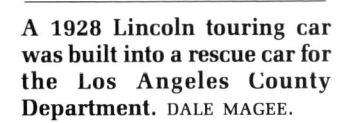

A 1928 Lincoln touring car was built into a rescue car for the Los Angeles County Department. DALE MAGEE.

Known as a "Little Giant," this truck was manufactured by the Chicago Pneumatic Tool Company, Chicago, Illinois, from 1912 to 1918. This 1912 Model H is powered by a four-cylinder Continental engine. The Newark (California) Volunteer Fire Department mounted the two chemical tanks and hose body. Many years later the rig was sold to Decota, California. Notice the gong and oil lamps. WAYNE SORENSEN.

Motorized Fire Apparatus of the West • 199

Locomobile trucks were made in Bridgeport, Connecticut, from 1912 until 1916. This much-rebuilt 1912 Locomobile tractor pulls a city service truck. This unit was first used at Richmond, California, then by the California State Hospital at Imola. PAUL DARRELL.

Aptos, California, used this 1917 Locomobile touring car to build their first apparatus. APTOS FIRE DEPARTMENT.

Marmon-Herrington built rugged off-road trucks and multiple-axle drive units for Fords. Here's a 1950 Marmon-Herrington airport foam truck, with American-LaFrance equipment, developed for the military. Pocatello, Idaho, acquired this rig from the United States Air Force. The foam pump is powered by a separate motor. Note three sweeping foam nozzles in front, below the bumper. WAYNE SORENSEN.

200 • Motorized Fire Apparatus of the West

Above: The Mason truck was introduced by William C. Durant. This rig is a 1923 Mason, with American-LaFrance chemical and hose equipment added by the American-LaFrance San Francisco service center, used by the Pleasanton (California) Fire Department. WAYNE SORENSEN.

Below: This 1912 Maxwell touring car was a fire patrol vehicle in California's Sierra National Forest. (The Maxwell's name was kept alive in the public's mind by comedian Jack Benny. Maxwell also built a truck. In 1921 it was the 10th most popular truck in terms of registrations throughout the United States.) FOREST SERVICE, U.S.D.A.

Motorized Fire Apparatus of the West • 201

The Moreland Motor Truck Company was located in Burbank, California, and produced one of the best-known West Coast trucks, from 1912 until 1941. Watt Moreland, the firm's founder, was both an industry spokesman as well as an important figure in the industrial development of Southern California. This 1915 Moreland chemical and hose car has been restored by the Burbank Fire Department. WAYNE SORENSEN.

This 1916 Moreland chassis was first used as a lumber truck by the Michigan-California Lumber Company at Camino, California. In 1924, the company made it into a fire engine by adding the small, front-mounted pump. Later, it was donated to the Placerville (California) Fire Department. Most recently, it has been seen parked outside an antique store in the Sierra foothills. WAYNE SORENSEN.

202 • *Motorized Fire Apparatus of the West*

Sonora, California, still has this 1921 Moreland combination chemical and hose car, and it is still run in parades. It's a Model 20-B, powered by a four-cylinder Continental engine. WAYNE SORENSEN.

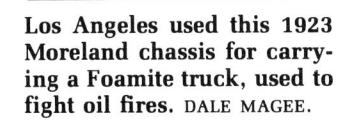

Los Angeles used this 1923 Moreland chassis for carrying a Foamite truck, used to fight oil fires. DALE MAGEE.

This 1927 Moreland (still with right-hand steering) is a combination chemical and hose wagon, which ran as part of a two-piece engine company, No. 25, in Los Angeles. GLEN ALTON.

Motorized Fire Apparatus of the West • 203

Above: Los Angeles Fire Department's Engine 39 tanker is mounted on a 1930 Moreland Roadrunner-7 chassis. The truck has a 650-gallon tank and a 150-gpm pump. Tankers served in brush areas and ran with engine companies. DALE MAGEE.

Below: Here is a 1930 Moreland American-LaFrance Type 36 two-tank chemical and hose car photographed at the San Francisco Exposition. The body had been added by the American-LaFrance service center in San Francisco. WAYNE SORENSEN.

Above: It must have been the Fourth of July when this picture was taken in Sandpoint, Idaho, where both the fire department and band look ready for the parade. The truck is a 1920 Nash chemical and hose, pulling a horse-drawn gasoline pumping engine. Nash, located in Kenosha, Wisconsin, was primarily a builder of autos, but made some trucks in this era. BONNER COUNTY HISTORICAL SOCIETY.

Below: The Packard Motor Car Company of Detroit manufactured some of the finest autos in the United States. They also built a truck from 1905-1923, which is recognizable because of its distinctive hood. This 1916 Model 36 Packard chemical and hose car was used by the Alameda (California) Fire Department. PAUL DARRELL.

Motorized Fire Apparatus of the West • 205

Above: Hayward, California's 1919 Model 36 Packard chemical and hose wagon has an overhead ladder rack. The tires are 36 x 5 rubber, mounted on Sewell cushion wheels, which had the outside solid rubber tire mounted on Sewell rubber flanges between side steel bands to help absorb road shocks. HAYWARD MUSEUM.

Below: Placerville, California, used this 1920 Packard combination chemical and hose car. It was outfitted with American-LaFrance equipment, at the American-LaFrance San Francisco factory service center. It had double chemical tanks. AMERICAN-LaFRANCE SERVICE CENTER.

206 • *Motorized Fire Apparatus of the West*

These days, Peterbilt is a well-known California truck. Production began in 1939, and the firm originated, more or less, out of the Fageol make of trucks. This 1956 Peterbilt COE (cab-over-engine) was used by the Fresno (California) Fire Department to build a 1,250-gpm quad. It had a Hale-ZMF pump and a Hall-Scott 935 engine. In its later life, the ladder racks were removed and a large, 2,600-gallon water tank installed to create a tanker. BOB ALLEN.

Pope-Hartford, of Hartford, Connecticut, built vehicles from 1906 until 1914. This photo, taken on January 13, 1912, shows the San Francisco Fire Department's first motorized apparatus test, held on the city's California Street hill. The 1912 Pope-Hartford in this picture passed all the tests. It was equipped with two 80-gallon chemical tanks, a reel carrying 300 feet of hose and ladders. It also had adapters for filling tanks off hydrants. The truck had a 40-hp, four-cylinder engine. WAYNE SORENSEN.

San Bernardino, California, acquired this 1912 Pope-Hartford chemical and hose car. Chains on the rear wheels were for traction on dirt roads. WAYNE SORENSEN.

Motorized Fire Apparatus of the West • 207

Above: Berkeley, California's 1912 Pope-Hartford chemical and hose car. The gas headlights and floodlight were supplied by a carbide generator. BERKELEY FIRE DEPARTMENT.

Below: This Pope-Hartford was a fireboat hose tender for the San Francisco Fire Department. It had a high-pressure Gorter battery and carried 2,000 feet of three-inch hose. Note the cab-over-engine design, common on early trucks. After San Francisco's 1989 earthquake, a fireboat had to pump water by hose to fire engines which then pumped it to the Marina district, where water mains had been broken. WAYNE SORENSEN.

Rambler automobiles were made by the Thomas B. Jeffery Company of Kenosha, Wisconsin (which later built the Jeffery "Quads" of World War I fame, and still later became part of Nash). Shown is one of two 1907 Ramblers, used as a chemical car by the Long Beach (California) Fire Department. A local machine shop converted the auto by adding two 40-gallon chemical tanks and a hose bed. LONG BEACH FIRE DEPARTMENT.

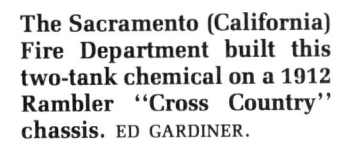

The Sacramento (California) Fire Department built this two-tank chemical on a 1912 Rambler "Cross Country" chassis. ED GARDINER.

The Reo Motor Car Company of Lansing, Michigan, was founded by Ransom E. Olds, who had also been the originator of the Oldsmobile auto. Reo built both autos and trucks. This is a 1913 Reo Model J chemical and hose car. Note the chain drive and tires. Holes in the solid rubber tires were to provide a cushioning effect. REO.

Motorized Fire Apparatus of the West • 209

This 1922 Reo Model F with hose and chemical equipment (by Hirsch) was used by the Huntington Park (California) Fire Department. (At the end of 1921, Reos were the third most popular truck registered in the United States.) WALT PITTMAN.

San Leandro, California, used this 1928 Reo Model BA chassis to carry a 150-gpm front-mounted pump and hose body. This rig has been restored. WAYNE SORENSEN.

For many years Reo named its product the "REO Speedwagon," a word known to only old-truck fanciers. In recent years a very popular rock music group called themselves the "REO Speedwagon," although pronunciation was the "R" "E" "O" "Speedwagon." The picture shows a 1931 REO Speedwagon, which the Fresno Fire Department shops built into a water tanker. LAVAL COMPANY, INC.

210 • Motorized Fire Apparatus of the West

This 1919 Republic truck was manufactured in Alma, Michigan. The chemical and hose car was Moscow, Idaho's first motorized apparatus. LATAH COUNTY HISTORICAL SOCIETY.

Star trucks were built by Durant Motors of Lansing, Michigan. This 1926 Star Six Compound Fleettruck, with a front-mounted 150-gpm pump and booster equipment, was used by the Pleasanton (California) Fire Department. PLEASANTON FIRE DEPARTMENT.

The truck is a 1930 Sterling, manufactured in West Allis, Wisconsin. It's pulling a trailer carrying a Cletrac bulldozer used in Los Angeles for fighting brush fires and keeping fire trails and fire breaks clear. BOB ALLEN.

Motorized Fire Apparatus of the West • 211

Stoddard-Dayton trucks were built by the Dayton Motor Company of Dayton, Ohio, for only two years, 1911 and 1912. The firm, however, built autos for a longer period of time. The picture shows a 1911 Stoddard-Dayton used by the Ogden (Utah) Fire Department as a chemical and chief's car. WAYNE SORENSEN.

The Menlo Park (California) Fire Department bought this 1917 one-ton Studebaker chassis (left) and outfitted it with two 40-gallon chemical tanks and other equipment from Seagrave. At right is 1916 Seagrave 350-gpm pumper. HAROLD SCHAPELHOUMAN.

Studebaker, of South Bend, Indiana, started out as a wagon builder, converting to the manufacture of autos and trucks at the turn of this century. Here, on one of their 1918 light truck chassis, is a chemical and hose car used by the Santa Cruz (California) Fire Department. ROBERT SAMS.

212 • Motorized Fire Apparatus of the West

A 1927 Studebaker Six carries a chemical and hose car for the Elk Valley (California) Fire Department. Note the similarity to passenger car styling. ROBERT SAMS.

Pacifica, California, used a 1928 Studebaker Big Six bus chassis for their rescue squad. JOHN GRAHAM.

Wallace, Idaho, created a city service truck by lengthening the chassis on this 1929 Studebaker President Eight auto chassis. Local firemen were paid additional money for after-hours work to build this truck. It exists today. DAN MARTIN.

This 1931 Studebaker President commercial chassis with an ambulance body was used by the Piedmont (California) Fire Department, one of the first in the state to offer ambulance service. PIEDMONT FIRE DEPARTMENT.

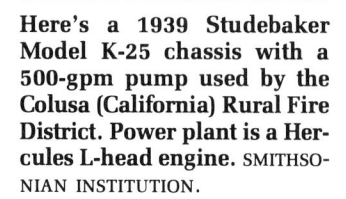

Here's a 1939 Studebaker Model K-25 chassis with a 500-gpm pump used by the Colusa (California) Rural Fire District. Power plant is a Hercules L-head engine. SMITHSONIAN INSTITUTION.

The Whitney Fire District, near Boise, Idaho, acquired this 1950 Studebaker 500-gpm pumper from a local dairy that had provided fire protection service. WAYNE SORENSEN.

214 • Motorized Fire Apparatus of the West

Tourist automobiles and light trucks were manufactured in Los Angeles from 1902 until 1910. They were designed by Watt Moreland (later associated with the truck bearing his name). This 1908 chemical and hose rig was delivered to Hollywood, and later became part of the equipment of the Los Angeles Fire Department. LOS ANGELES FIRE DEPARTMENT.

This Velie, made in Moline, Illinois, was used by the Fire Dispatch Patrol System of Oakland, California. (Dispatch patrols were owned by insurance companies, and they would try to protect property — such as retail store inventories — from water damage.) Later, this rig became a hose wagon for the Oakland Fire Department. WAYNE SORENSEN.

Motorized Fire Apparatus of the West • 215

Above: White is one of the country's best-known makes of trucks. The firm started in Cleveland, Ohio, about the turn of this century and remained there until recently, when it became affiliated with Volvo and moved most of its United States operations to South Carolina. In World War I, the White firm received a large order from the United States Army for fire engines. Also, for a brief time, they marketed their own fire apparatus, which utilized a Northern pump. Shown is a 1911 White chemical car, used in Livingston, Montana. Note the right-hand drive and windshield. WAYNE SORENSEN.

Below: The first motorized equipment in Chico, California, was this 1912 White chemical and hose car. It was so difficult to start that a raised wooden platform was built inside the fire hall, on to which the vehicle would be backed while being parked. This would give it a rolling start when the bell rang. ROBERT SAMS.

216 • Motorized Fire Apparatus of the West

Above: Tacoma, Washington, used this 1915 White chemical and hose wagon. RALPH DECKER.
Below: The Oregon Short Line Division of the Union Pacific Railroad used this 1918 White chemical and hose car to protect their property. The rig was rebuilt several times. WAYNE SORENSEN.

Above: Fresno, California, firemen could feel every bump as they responded in this 1918 White double-tank chemical car. The acetylene tank on the running board was for headlamps and floodlight. WAYNE SORENSEN COLLECTION.

Below: Bountiful, Utah, restored this 1919 White, which originally belonged to Buhl, Idaho. It has a Northern 250-gpm pump. The original rig had solid rubber tires. As of December 31, 1921, Whites were the fifth most popular truck registered in this country. WAYNE SORENSEN.

Above: Flagstaff, Arizona, used this 1923 White 350-gpm pumper for many years. ROLAND BOULET COLLECTION.
Below: A 1930 White Model 61 chassis was used to build this 500-gpm pumper for the Los Angeles County Fire Department. HUNTINGTON LIBRARY.

Motorized Fire Apparatus of the West • 219

Above: This 1930 White Model 63 was a city service truck for Grand Island, Nebraska. STUHR MUSEUM OF THE PRAIRIE PIONEER, GRAND ISLAND.

Below: On the right is a 1931 White tractor that pulled a 1908 American-LaFrance 75-foot aerial for the San Jose (California) Fire Department. It's shown as it is being replaced by a 1942 Seagrave 65-foot aerial. WAYNE SORENSEN.

220 • *Motorized Fire Apparatus of the West*

Above: **During World War II, our civil defense effort included training civilians as auxiliary fire fighters. This 1943 photo shows several in San Francisco training with a White Horse (trade name for a White step van). To the left is a FABCO.** VOLVO/WHITE.

Below: **The Willys-Overland Company of Toledo, Ohio, built the chassis for this early 1920s chemical and hose car used by Belvedere, California.** ROBERT SAMS.

Motorized Fire Apparatus of the West • 221

Above: In 1910, this Winton runabout (an auto) was donated to the Piedmont (California) Fire Department to pull their hose cart. A few years later it was converted into a chemical and hose car. PIEDMONT FIRE DEPARTMENT.

Below: The Wisconsin truck was built in Baraboo, and later in Loganville, Wisconsin. The Spokane (Washington) Fire Department's shop used this 1914 Wisconsin chassis to build a hose wagon. DICK ADELMAN.

222 • Motorized Fire Apparatus of the West

This 1923 Wisconsin chassis was built into a city service truck by the shops of the Spokane (Washington) Fire Department. DICK ADELMAN.

Spokane, Washington, used this 1923 Wisconsin double-tank chemical. It had right-hand drive. DICK ADELMAN.

Above: This American-LaFrance factory shows a 1925 400 gpm pumper purchased by Elsinore, California. Both the headlights and the floodlight were 12 inches in diameter, and there was a 40-gallon chemical tank. AMERICAN-LAFRANCE.

Below: This Seagrave factory photo shows a 1934 rescue squad delivered to Bakersfield, California. Equipment included floodlights and a booster tank. SEAGRAVE.

224 • Motorized Fire Apparatus of the West

Above: **This is a 1914 Christie tractor, used by Pasadena, California, to pull a 1888 Ahrens steam pump. Rig has 1923 license plates.** DALE MAGEE.

Below: **A 1929 Mack Tractor, used by the San Francisco Fire Department to pull a city service trailer built in the department's shops. Rig presently belongs to the St. Francis Hook and Ladder Society.** SAN FRANCISCO FIRE DEPARTMENT.

Motorized Fire Apparatus of the West • 225

A multiple alarm fire in San Francisco in the mid-1930s. A 1928 Mack with a 35-foot Gorter water tower has just shut down. Next to the water tower, we see the top of a city service trailer; toward the curb are a White ambulance and Buick chief's buggy. Behind them is a 1928 Kleiber light truck. At the intersection we see the front of a 1932 American-LaFrance pumper, parked next to a 1926 White hose tender. Rigs beyond the intersection cannot be identified. WAYNE SORENSEN COLLECTION.

Bibliography

Ahrens-Fox catalogs, circa 1905-1918.

"American Cars Since 1775, The." *Automobile Quarterly, Inc.,* New York: 1971.

American-LaFrance catalogs, a 1912-1925.

American-LaFrance Motor Fire Apparatus Instruction Manual, 3rd ed. Elmira, New York: The Company, 1921.

Association of Automobile Manufacturers. *Handbook of Automobiles, 1925-1926.* New York: Dover, 1970.

Association of Licensed Automobile Manufacturers. *Handbook of Automobiles, 1915-1916.* New York: Dover, 1970.

Bell, Doug. *Photo Album, Early Chevrolet History, Trucks — 1918-1945.* Los Angeles: Clymer Publications, 1966.

Birchfield, Rodger. *Stutz Fire Engine Company, A Pictorial History.* Indianapolis: the author, 1978.

----------. *New Stutz Fire Apparatus Co., Inc.* Indianapolis: the author, 1981.

James Boyd & Bro. catalog, circa 1910.

Burness, Tad. *American Truck Spotter's Guide, 1920-1970.* Osceola, WI: Motorbooks International, 1978.

Burgess-Wise, David. *Fire Engines & Fire Fighting.* Norwalk, CT: Longmeadow Press, 1977.

Burks, John. *Working Fire — The San Francisco Fire Department.* Mill Valley, CA: Squarebooks, 1982.

Crismon, Fred W. *U.S. Military Wheeled Vehicles.* Sarasota, FL: Crestline, 1983.

Dammann, George. *Sixty Years of Chevrolet.* Glen Ellyn, IL: Crestline, 1972.

----------. *Illustrated History of Ford.* Glen Ellyn, IL: Crestline, 1970.

Decker, Ralph and Clyde Talbert. *100 Years of Fire Fighting in the City of Destiny: Tacoma, Washington.* Seattle: Grange Printing, 1981.

Ditzel, Paul C. *Fire Engines, Fire Fighters.* New York: Rutledge Books, 1976.

---------- and the editors of Consumers Guide. *The Complete Book of Fire Engines.* New York: Crown Publishers, 1982.

Fire Apparatus Photo Album of the American-LaFrance 150th Anniversary. Naperville, IL: The Visiting Fireman, 1982.

Fire Apparatus Photo Album of the Greenfield Village Musters. Naperville, IL: The Visiting Fireman, 1984.

Fire Apparatus Photo Album of the Valhalla Musters. Naperville, IL: The Visiting Fireman, 1983.

Frady, Steven R. *Red Shirts and Leather Helmets, Volunteer Fire Fighting on the Comstock Lode.* Reno: University of Nevada Press, 1984.

FWD catalog, circa 1918.

Georgano, G.N., editor. *The Complete Encyclopedia of Commercial Vehicles.* Iola, WI: Krause Publications, 1979.

Goodenough, Simon. *Fire, The Story of the Fire Engine.* Secaucus, NJ: Chartwell Books, 1978.

B.F. Goodrich Co. *Motor Trucks of America, Vol. IV, 1916.* Akron: the Company, 1916.

Hagy, Steve. *Howe Fire Apparatus Album.* Naperville, IL: The Visiting Fireman, 1984.

Hall, Asa E. and Richard Langworth. *The Studebaker Company: A National Heritage.* Contoocook, NH: Dragonwyck, 1983.

Hart, Arthur A. *Fighting Fires on the Frontier.* Boise, ID: Boise Fire Department Association, 1976.

Hass, Ed. *Ahrens-Fox, The Rolls Royce of Fire Engines.* Sunnyvale, CA: the author, 1982.

Heiser, Harvey Michael. *Memories of the Fire Service.* Sacramento: the author, 1941.

Ingram, Arthur. *Fire Engines in Color.* New York: Macmillan, 1973.

Karolevitz, Robert F. *This Was Trucking.* Seattle: Superior, 1966.

Kenoyer, Natlee. *The Fire Horses of San Francisco.* Los Angeles: Westernlore Press, 1970.

Kimes, Beverly Rae. *Standard Catalog of American Cars.* Iola, WI: Krause Publications, 1985.

King, William T. *History of the American Fire Engine.* Chicago: Owen Davies, 1960.

Klass, George. *Fire Apparatus: A Pictorial History of the Los Angeles Fire Department.* Inglewood, CA: Mead, 1974.

LaFrance Fire Engine Company catalog, circa 1900.

Laval, Jerome with Stan Benson. *Second Alarm, The Fresno Fire Department.* Fresno, CA: Graphic Technology, 1976.

Los Angeles County Fire Department, The Story of a Fire Department. Los Angeles: The Los Angeles County Firemen's Benefit & Welfare Association, 1975.

Mack (International Motor Co.) catalogs, circa 1915-1918.

Matches, Alex. *It Began with a Ronald.* Vancouver, BC: the author, 1974.

McCall, Walter. *American Fire Engines Since 1900.* Glen Ellyn, IL: Crestline, 1976.

McNeish, Robert H. *The Automobile Fire Apparatus Operator.* Los Angeles: The Times-Mirror Press, 1926.

McPherson, Thomas A. *The Dodge Story.* Glen Ellyn, IL: Crestline, 1975.

Miller, Dennis. *The Illustrated Encyclopedia of Trucks and Buses.* New York: Mayflower, 1982.

Montville, John B. *Bulldog, The World's Most Famous Truck.* Tucson, AZ: Aztec, 1979.

----------. *Mack.* Newfoundland, NJ: Haessner, 1973.

Nailen, Richard L. *Guardians of the Garden City, The History of the San Jose Fire Department.* San Jose: Smith & McKay, 1972.

Page, Victor W. *The Modern Motor Truck.* New York: Henley, 1921.

Peckham, John M. *Fighting Fire with Fire.* Newfoundland, NJ: Haessner, 1972.

Rice, Gini. *Relics of the Road #1, GMC Gems, 1900-1951.* New York: Hastings House, 1971.

----------. *Relics of the Road # 2, Keen Kenworth Trucks, 1915-1955.* New York: Hastings House, 1973.

------. *Relics of The Road # 3, Impressive Internationals.* Lake Oswego, OR: Truck Tracks, 1975.

Salt Lake City Firemen's Relief Association. *A Pictorial History of the Salt Lake City Fire Department, 1871-1976.* Salt Lake City: the Association, 1976.

Schwalbe, Timothy. *Fire Apparatus Photo Album of the Crash-Rescue Units.* Naperville, IL: The Visiting Fireman, 1985.

Seagrave Motor Fire Apparatus Text Book. Columbus, OH: The Company, 1919.

Sorensen, Lorin. *The Commercial Fords.* St. Helena, CA: Silverado Publishing Co., 1984.

Stern, Philip Van Doren. *Tin Lizzie, The Story of the Fabulous Model T Ford.* New York: Simon & Schuster, 1955.

Sytsma, John F. *Ahrens-Fox Album.* Medina, OH: the author, 1973.

----------. *Ahrens-Fox Factory Registration Numbers of Motorized Fire Apparatus, 1912-1957.* Medina, OH: the author, 1974.

----------. and Robert Sams. *Ahrens-Fox, A Pictorial Tribute to a Great Name in Fire Apparatus.* Medina, OH: the author, 1971.

Troyer, Howard William. *The Four Wheel Drive Story.* New York: McGraw-Hill, 1954.

Vanderveen, Bart H., editor. *Fire & Crash Vehicles from 1950.* New York: Frederick Warne & Co., 1976.

----------, editor. *Fire Fighting Vehicles, 1840-1950.* New York: Frederick Warne & Co., 1972.

Wagner, James K. *Ford Trucks Since 1905.* Glen Ellyn, IL: Crestline, 1978.

Waterous Engine Works catalog, circa 1913.

Webb catalogs, circa 1910-1913.

Wren, James A. and Genevieve J. *Motor Trucks of America.* Ann Arbor: The University of Michigan Press, 1979.

This picture, from about 1930, shows three Studebakers. Banner calls them "Studebaker Mountain Fire Engines," and indicates that they were sold by a dealer in San Bernardino, California. Lettering on tool compartment appears to say: "Crestline District, San Bernardino Mountains." MOTOR VEHICLE MANUFACTURERS ASSOCIATION.

Index

Manufacturers

— A —

Ahrens 231
Ahrens-Fox 13, 15, 71, 90-96, 169
American Car & Foundry (ACF).... 21-23, 59
American Fire Apparatus Co. 24-27
American-LaFrance.... 10, 14, 15, 17-19, 75, 92, 97-109, 124, 141, 146, 162, 166, 171, 176, 177, 201, 204, 206, 220, 224, 227
American-Marsh 25
Amoskeag 10
Atterbury 135
Autocar 32, 49, 76, 160, 161
Available 55

— B —

Barton 24, 25, 79, 121, 171
Bean, John 9, 50-52, 178
Bean-Rool 169
Berkeley Pump Co. 77
Biederman 162
Boardman 20, 21, 27
Boyd 18, 21
Brill, J.G. 23
Brockway 18, 107, 108, 162
Buckeye Manufacturing Co. 118, 119
Buda 100
Budd-Michelin 186
Buffalo 103
Buick 8, 9, 99, 140, 159, 163, 224
Byron-Jackson 40, 126-128

— C —

Cadillac 164, 165, 199
Central 26
Chadwich Engineering Works 73
Challenger 28-32
Champion 113-115
Chevrolet 15, 25-27, 33, 49, 51, 64, 71, 81, 113, 114, 117, 149, 165-168
Chicago Pneumatic Tool Co. 199
Childs, O.J. 8, 9
Christie 112, 113
Cletrac 211
Coast 32-37, 47, 119, 150
Coleman 168
Continental 136, 147, 155, 157, 158, 199, 203
Cooney, P.J. Co. 73
Corbitt 169
Couple-Gear 169
Crosley 169
Crown 38-41, 133
Curtis 42-45, 133
Curtis-Heiser 45, 230

— D —

Darley 78, 113-115
Dayton Motor Co. 212
Defiance 119
Delco 164
DeMartini 170

Diamond-T ... 15, 30, 31, 48, 50, 65, 73, 83, 170, 171
Dodge 15, 18, 25-27, 47, 48, 50, 52, 55, 60, 68, 75, 77, 78, 80, 81, 129, 171-173, 185
Dorris 174
Duplex 56, 119, 120
Durant 211

— E —

Evinrude 27, 29

— F —

FABCO 6, 33, 46-49, 182, 221
Fageol 21-23, 28, 29, 46-49, 86, 174, 175, 207
Federal 32, 55, 136, 176
Figgie, International, Inc................ 99
Fire Truck, Inc. (FTI) 53
Flxible 176
Foamite 203
Food Machinery Corp.,
 (FMC) 50-52
Ford 8, 9, 15, 17, 20, 21, 25-27, 32, 33, 37, 42, 43, 46, 47, 61, 69, 71, 78, 79, 84, 88, 107, 114, 115, 118, 119, 134, 150, 158, 166, 177-182
Frank and St. Gem Fire Extinguisher
 Company 53
Fresno F.D. shops 176, 182, 183
FWD 69, 115-117

— G —

Gamewell 145
General 27, 53-56
GMC..... 5, 27, 35, 42-44, 50, 52, 59, 87-89, 99, 108, 109, 139, 140, 176, 183-185
Gorham 9, 18, 56-59
Gorter 126, 198, 208, 224
Graham Brothers 69, 75, 185-187
Graham-Paige 19
Gramm 119, 187
Gramm-Bernstein 63
Gramm-Kincaid 63
Grey Fire Apparatus 61
Grove 150
Grumman Emergency Products 119

— H —

Hale....... 31, 36, 60, 61, 78-80, 82, 83, 96, 111, 117, 132, 133, 137, 154, 155, 158, 207
Hall-Scott...... 9, 21, 23, 28, 29, 31, 33, 34, 36, 39, 45-47, 58, 59, 82-84, 86, 101, 117, 128, 129, 132, 140, 149, 153, 175, 183, 196, 207
Harley-Davidson 187
Hedberg 60, 61
Heiser 43, 45, 230
Hercules 111, 176, 181, 214
Herschall-Spillman 91
Hewitt-Ludlow 188
Hi-Ranger 84
Hirsch 210

Holmes 161
Howard-Cooper 61-65, 101, 185
Howe 12, 118-121
Hudson 188
Hug 189

— I —

Indiana 38, 76
International 25, 27, 33-38, 42, 44, 51, 64, 68, 69, 78, 80, 83, 84, 114, 120, 189-191, 199, 230

— J —

Jeep 121
Jeffery 192, 209

— K —

Kelly 193
Kelly-Springfield 192, 193
Kelsey-Hayes 47
Kenney 75
Kenworth 45, 54, 80, 82, 87, 101, 193-196
Kissel 197
Kleiber 5, 198, 224
Knox 22, 122-124
Knox-Martin 18, 123, 124

— L —

Lambert 119
Larkin 75
Larson & Krog 182
Lincoln 199
Little Giant 199
Locomobile 18, 27, 29, 200
Lutweiler 11
Luverne 67
Lycoming 103, 156

— M —

Mack 61, 91, 124-133, 140, 224, 231
Marmon-Herrington 200
Mars 148
Mason 10, 201
Maxim 43, 133, 134
Maxwell 201
Menco 84
Metro 113
Moore 92
Moreland 202-204, 215
Morse 32, 117, 132

— N —

Nash 135, 205
National Belting & Hose Co. 53
Northern 62, 66-69, 153, 218
Nott 69, 70, 112

— O —

Oberdorfer 64, 185
Oldsmobile 159
Oren Roanoke Corp 119

Motorized Fire Apparatus of the West • **229**

Oshkosh . . . 83

— P —

Packard . . . 18, 54, 69, 205, 206
Paige . . . 18
Peterbilt . . . 23, 29, 30, 31, 33, 36, 82, 153, 183
Pierce . . . 39
Pierce-Arrow . . . 27, 29, 53
Pirsch . . . 135-141, 194
Pitman . . . 41, 191
Pope-Hartford . . . 207, 208
Proctor-Keefe . . . 172
Pyrene . . . 190

— Q —

— R —

Ralston Iron Works . . . 188
Rambler . . . 192, 209
Redwine Motor Co. . . . 181
Reo . . . 15, 63, 76, 149, 209, 210
Republic . . . 211
Robinson . . . 72, 73
Roney . . . 73
Roto-Rays . . . 148

— S —

Schacht . . . 91, 95
Schneer . . . 74, 75
Seagrave . . . 7, 11, 16, 18, 41, 56-59, 64, 89, 92, 101, 104, 117, 123, 125, 148, 188, 196, 212, 220, 227
Selden . . . 8, 9
Sewell . . . 206
Simplex . . . 97
Snorkel . . . 41
Spicer . . . 40
Standard-Webb . . . 7
Star . . . 211
Stemple Fire Extinguisher Co. . . . 73
Sterling . . . 33, 34, 120, 211
Stoddard-Dayton . . . 212
Studebaker . . . 53, 119, 212-214, 226
Stutz . . . 151-153

— T —

Thomas . . . 133, 159
Tourist . . . 215

— U —

Union Iron Works . . . 126
United Fire Equipment Co. . . . 194, 196
U.S.A. . . . 154-156

— V —

Van Pelt . . . 46, 75-84, 183
Velie . . . 215
Viking . . . 15
Volvo/White . . . 161, 216

— W —

Walter . . . 89
Ward-LaFrance . . . 157, 158
Waterous . . . 12, 18, 32-35, 37, 39, 47, 85, 86, 116, 117, 119, 158
Waukesha . . . 28, 55, 136, 138
Webb . . . 7, 18, 87, 159
Wentworth and Irwin . . . 86, 87
Western States Fire Apparatus Inc. . . . 88, 89
Westland Trailers . . . 73
White . . . 30, 33, 34, 38, 48, 77, 79, 81, 82, 113, 141, 146, 194, 216-221, 224
Willys . . . 121
Willys-Overland . . . 221
Winton . . . 222
Wisconsin . . . 222, 223
Wisconsin Motors . . . 91, 108

— X —

— Y —

Yankee . . . 89

— Z —

Cities and Places

— A —

Aberdeen, WA . . . 93, 135
Ada, OK . . . 57
Alameda, CA . . . 18, 46, 85, 86, 191, 205
Allentown, PA . . . 125, 159
Alma, MI . . . 211
Almeda, ID . . . 118
American Fork, UT . . . 40, 68
Anaheim, CA . . . 7, 137
Anchorage, AK . . . 17, 115, 120, 158
Anderson, IN . . . 119
Aptos, CA . . . 44, 81, 200
Arcata, CA . . . 48
Arden, CA . . . 169
Ardmore, PA . . . 161
Ashland, OR . . . 63
Atwater, CA . . . 12, 107, 165
Auburn, CA . . . 32, 134, 163
Azusa, CA . . . 40

— B —

Bakersfield, CA . . . 57, 58, 187, 227
Baraboo, WI . . . 222
Battle Creek, MI . . . 25
Battle Mountain, NV . . . 43, 111
Beaverton, OR . . . 94, 113
Belvedere, CA . . . 221
Ben Lomond, CA . . . 37, 81
Bend, OR . . . 136, 176
Benecia, CA . . . 55, 114, 117
Benton County, WA . . . 88
Berkeley, CA . . . 7, 18, 21-23, 57, 59, 98, 123, 126, 137, 151, 186, 208
Bethel, WA . . . 65
Beverly Hills, CA . . . 94, 117, 194
Bluefield, VA . . . 99
Boise, ID . . . 114, 119, 125, 142, 214
Bountiful, UT . . . 218
Brantford, Ont. . . . 85
Bridgeport, CT . . . 200
Brisbane, CA . . . 173
Buffalo, NY . . . 111
Burbank, CA . . . 107, 128, 202
Burlingame, CA . . . 168
Butte, MT . . . 185, 197

— C —

Caldwell, ID . . . 159
Camino, CA . . . 202
Campbell, CA . . . 178
Carmel, CA . . . 67, 129
Carmel Highlands, CA . . . 80
Carson, NV . . . 147
Carson City, NV . . . 144
Castro Valley, CA . . . 158
Castroville, CA . . . 82
Cedar Falls, IA . . . 15
Chatsworth, CA . . . 165
Chicago, IL . . . 57, 65, 113, 117, 170, 199
Chico, CA . . . 216
Cincinnati, OH . . . 91, 162
Citrus Heights, CA . . . 32, 150, 174
Clackamas County, WA . . . 64
Clear Lake, CA . . . 50, 52, 117
Cleveland, OH . . . 216
Clintonville, WI . . . 115, 141, 143
Cloverdale, CA . . . 52
Colorado Springs, CO . . . 187
Colton, CA . . . 57, 58
Columbia, CA . . . 172
Columbus, OH . . . 56, 57, 141
Colusa, CA . . . 162, 214
Conshoncken, PA . . . 117
Cornelius, OR . . . 89
Coronado, CA . . . 128
Cortland, NY . . . 107, 162
Creston, IA . . . 135
Crockett, CA . . . 32, 84

— D —

Danville, CA . . . 76, 156
Davenport, CA . . . 167
Davis, CA . . . 46, 47
Dayton, OH . . . 212
Decota, CA . . . 199
Deer Lodge, MT . . . 177
Del Paso-Roble, CA . . . 31
Denver, CO . . . 69, 106, 113, 123, 148, 152, 169
Detroit, MI . . . 53, 141, 172, 175, 205
Diamond Springs, CA . . . 50, 158
Dillon, CO . . . 120
Dixon, CA . . . 47
Douglas, AZ . . . 145

— E —

Eagle, ID . . . 119
El Cerrito, CA . . . 110
El Segundo, CA . . . 195, 196
Elk Grove, CA . . . 75, 77, 213
Elko, NV . . . 98, 191
Elmira, CA . . . 171
Elmira, NY . . . 107, 157, 158
Elsinore, CA . . . 227
Emeryville, CA . . . 35, 47, 159
Ephraim, UT . . . 42
Estacada, OR . . . 64
Eureka, CA . . . 95

— F —

Fairfax, CA . . . 78
Fairfield, CA . . . 12
Flagstaff, AZ . . . 219
Fort Collins, CO . . . 155, 166
Fort Holabird, MD . . . 155
Freedom, CA . . . 34, 39
Fremont, CA . . . 30, 163
Fresno, CA . . . 99, 113, 176, 182, 183, 207, 210, 218

— G —

Gallup, NM . . . 172
Galt, CA . . . 81
Gaston, OR . . . 26
Gila National Forest, NM . . . 187
Glendale, AZ . . . 135
Glendora, CA . . . 100
Glenn Ellen, CA . . . 185
Glenns Ferry, ID . . . 62, 133
Globe, AZ . . . 72
Gold Beach, OR . . . 88
Grand Island, NE . . . 220
Grand Junction, CO . . . 144
Grand Rapids, MI . . . 169
Grants Pass, OR . . . 36, 37

— H —

Hammett, ID . . . 50
Hartford, CT . . . 207
Hartford, WI . . . 197
Hayward, CA . . . 206
Hemet, CA . . . 109
Henderson, NC . . . 169
Higginwood, CA . . . 29, 31
Highland, IL . . . 189
Hillsboro, OR . . . 151
Hillsborough, CA . . . 171
Hoboken, NJ . . . 113
Hollywood, CA . . . 215
Homestead, CA . . . 188
Horseshoe Bend, ID . . . 155
Huntington Park, CA . . . 210

— I —

Idaho City, ID . . . 121
Imola, CA . . . 200
Imperial Palace Hotel . . . 58
Indianapolis, IN . . . 119, 151
Iola, WI . . . 161

— J —

Jackson, MI . . . 25
Jerome, ID . . . 111

230 • *Motorized Fire Apparatus of the West*

Joplin, MO 159

— K —

Kansas City, MO 61
Keizer, OR 140
Kenosha, WI 135, 192, 205, 209
Kentfield, CA 102, 141
Kern County, CA 82, 160, 161
King County, WA 65, 230
Kingsburg, CA 108

— L —

Lake Grove, OR 56
Lake Tahoe, CA 27
Lakeport, CA 44
Lansing, MI 159, 209, 211
Larkspur, CA 180
Larue, OH 119
Las Vegas, NV 58, 149
Leadville, CO 118
Lebanon, OR 189
Leggett, CA 108
Lekkogg, ID 130
Lewiston, ID 61, 136
Lima, OH 62, 63
Littleton, CO 168, 175
Livermore, CA 71
Livingston, MT 216
Lodi, CA 185
Logan, UT 18, 98
Loganville, WI 222
Long Beach, CA 57, 72, 93, 96, 144, 209
Los Altos, CA 180
Los Angeles, CA 33, 38-41, 53, 55, 57, 69, 71, 104, 107, 112, 113, 117, 126, 132, 140, 148, 152, 153, 177, 179, 194, 196, 199, 203, 204, 211, 215, 219
Los Banos, CA 102
Los Gatos, CA 51
Loundonville, OH 176
Lowry Field, CO 123
Luverne, MN 67

— M —

Madera, CA 57
Madison, MO 174
Magalia, CA 77, 178
Manchester, MO 27
Mare Island, CA 105
Marin County, CA 51
Martinez, CA 33
Marysville, WA 55
Mason City, IA 57
Menlo Park, CA 95, 212
Middleboro, MA 133
Middletown, NY 57
Mill Valley, CA 170
Milwaukie, OR 61
Minneapolis, MN 69, 125
Mission San Jose, CA 163
Modesto, CA 76, 80, 83
Moffit Field, CA 169
Moline, IL 215
Monterey, CA 38
Montreal, Que. 57
Moscow, ID 211
Mountain View, CA 84, 181

— N —

Nampa, ID 103, 152, 164, 175
Napa, CA 34, 57, 143
New London, CT 10
New York, NY 113, 117
Newark, CA 199
Niles, CA 48

— O —

Oakdale, CA 51, 75-79
Oakland, CA 12, 21, 28, 32, 33, 36, 43, 45, 53, 56-58, 69, 89, 92, 100, 108, 113, 139, 140, 156, 164, 174, 184, 192, 215

Ogden, UT 212
Oklahoma City, OK 27
Ophir Hills, CA 48
Orland, CA 133, 179
Oroville, CA 97
Owyhee, NV 131

— P —

Pacific Grove, CA 38
Pacifica, CA 213
Palo Alto, CA 78, 83
Pasadena, CA 57, 142, 170, 232
Payette, ID 132, 150
Petaluma, CA 69-71, 103, 190
Philadelphia, PA 23, 73
Phoenix, AZ 130
Piedmont, CA 49, 184, 188, 214, 222
Placerville, CA 27, 202, 206
Pleasant Hill, CA 161
Pleasant Valley, CA 26
Pleasanton, CA 164, 201, 211
Pocatello. ID 19, 100, 116, 131, 147, 200
Point Reyes, CA 51
Pollock Pines, CA 120, 167, 173
Port Angeles, WA 149
Portland, OR 62, 63, 73, 86, 87, 94, 97, 101, 113, 134, 177, 189
Presidio of Monterey, CA 162
Presidio of San Francisco, CA 154

— Q —

Quincy, CA 29, 173

— R —

Rawlins, WY 102
Redwood, CA 74, 134
Regina, Sask. 57
Reno, NV 105
Renton, WA 62, 63
Richmond, CA 79, 138, 145, 163, 200
Rincon Valley, CA 121
Rio Nido, CA 166
Rio Vista, CA 157, 171
Ripon, CA 165
Riverside, CA 141
Rochester, NY 11
Rocky Mountain National Park, CO 154
Ross, CA 127
Roswell, NM 54
Rupert, ID 148

— S —

Sacramento, CA .. 29, 42, 57, 58, 90-92, 138, 181, 209
Salida, CA 129
Salida, CO 197
Salinas, CA 99, 199
Salt Lake City, UT 35, 43, 96, 115, 143
San Antonio, TX 189
San Bernardino, CA 207, 226
San Diego, CA 57, 142
San Francisco, CA 5, 46, 47, 57, 71, 74, 75, 89, 92, 93, 99, 100, 106, 108, 113, 124, 126, 157, 166, 171, 181, 186, 188, 197, 198, 201, 204, 206-208, 221, 224, 232
San Jose, CA 35-37, 51, 57, 59, 61, 99, 103, 105, 124, 153, 183, 220
San Juan Bautista, CA 79
San Leandro, CA 210
San Marino, CA 130
San Pablo, CA 34
San Quentin Prison, CA 108
San Rafael, CA 83, 104
Sand Point, ID 205
Santa Barbara, CA 54, 143, 146
Santa Clara, CA 145
Santa Clara County, CA 82
Santa Cruz, CA 167, 212
Sausalito, CA 188
Scotts Valley, CA 43
Seattle, WA 43, 45, 57, 62, 63, 86, 100, 113, 126-128, 132, 141, 146, 192, 193

Shelton, OR 63
Sierra National Forest, CA 201
Sonora, CA 203
Soquel, CA 80
South Bend, IN 212
South Lake Tahoe, CA 100
South Pasadena, CA 57
South St. Paul, MN 85
Spokane, WA 62, 135, 175, 194, 222, 223
Spreckels, CA 76
Springfield, MA 123
Springfield, OH 51, 192
St. Anthony, ID 68
St. Helena, CA 74, 75
St. Louis, MO 53, 73, 87, 174
St. Paul, MN 12, 85
Stanford, University 78
Staunton, VA 73
Stockton, CA 57, 129, 131
Summit, WA 73
Sumner, WA 193

— T —

Tacoma, WA 57, 112, 113, 139, 141, 217
Thermopolis, WY 96
Tipton, IN 51, 79
Toledo, OH 221
Tremonton, UT 33
Truckee, CA 28
Tucson, AZ 94
Tulare, CA 186
Tulsa, OK 60, 61
Tumwater, WA 45
Turlock, CA 12, 118

— U —

Union City, CA 78
Universal City Studios 55
Utica, NY 8, 9

— V —

Vallejo, CA 110, 122
Vancouver, BC 10, 141
Vernon, CA 84
Victoria, BC 45, 69
Vincennes, IN 159
Visalia, CA 57

— W —

Wallace, ID 213
Walnut Creek, CA 77, 163
Walnut Grove, CA 49
Walsenberg, CO 149
Waterford, CA 77
Watsonville, CA 71
Wayne, PA 85, 117
Weiser, ID 54
West Allis, WI 211
West Covina, CA 39
Westminster, CO 193
Wichita Falls, TX 57
Willow Oak Park, CA 178
Willows, CA 123, 124
Windsor, Ont. 57
Winlock, WA 64
Winona, MN 56
Winslow, AZ 150
Winters, CA 110
Woodbridge, CA 185
Woodside, CA 129
Worthington, MN 67

— X —
— Y —

Yerington, NV 42, 136, 138

— Z —

The George Heiser Body Company of Seattle, working in conjunction with L.N. Curtis, outfitted this 1956 International pumper for use by the King County (Washington) Fire District. GEORGE HEISER BODY COMPANY, INC.